T0038858

A NOTE ON THE AUTHOR

Grace Lindsay is an Assistant Professor of Psychology and Data Science at New York University. After completing her PhD in 2018 at the Center for Theoretical Neuroscience at Columbia University, she went on to a postdoctoral fellowship at University College London, where her research focused on building mathematical models exploring sensory processing. Before that, she earned a bachelor's degree in Neuroscience from the University of Pittsburgh, and received a research fellowship to study at the Bernstein Center for Computational Neuroscience in Freiburg, Germany.

She was awarded a Google PhD Fellowship in Computational Neuroscience in 2016 and has spoken at several international conferences. She lives in New York City with her husband and two children.

@neurograce

Some other titles in the Bloomsbury Sigma series:

Sex on Earth by Jules Howard
Spirals in Time by Helen Scales
A is for Arsenic by Kathryn Harkup
Suspicious Minds by Rob Brotherton
Herding Hemingway's Cats by Kat Arney
The Tyrannosaur Chronicles by David Hone
Soccermatics by David Sumpter
Wonders Beyond Numbers by Johnny Ball
The Planet Factory by Elizabeth Tasker
Seeds of Science by Mark Lynas
The Science of Sin by Jack Lewis
Turned On by Kate Devlin
We Need to Talk About Love by Laura Mucha
Borrowed Time by Sue Armstrong
The Vinyl Frontier by Jonathan Scott
Clearing the Air by Tim Smedley
The Contact Paradox by Keith Cooper
Life Changing by Helen Pilcher
Sway by Pragya Agarwal
Kindred by Rebecca Wragg Sykes
Our Only Home by His Holiness The Dalai Lama
First Light by Emma Chapman
Models of the Mind by Grace Lindsay
The Brilliant Abyss by Helen Scales
Overloaded by Ginny Smith
Handmade by Anna Ploszajski
Beasts Before Us by Elsa Panciroli
Our Biggest Experiment by Alice Bell
Worlds in Shadow by Patrick Nunn
Aesop's Animals by Jo Wimpenny
Fire and Ice by Natalie Starkey
Sticky by Laurie Winkless
Racing Green by Kit Chapman

MODELS OF THE MIND

HOW PHYSICS, ENGINEERING AND MATHEMATICS HAVE SHAPED OUR UNDERSTANDING OF THE BRAIN

Grace Lindsay

BLOOMSBURY SIGMA
LONDON · OXFORD · NEW YORK · NEW DELHI · SYDNEY

In memory of my father

BLOOMSBURY SIGMA
Bloomsbury Publishing Plc
50 Bedford Square, London, WC1B 3DP, UK
29 Earlsfort Terrace, Dublin 2, Ireland

BLOOMSBURY, BLOOMSBURY SIGMA and the Bloomsbury Sigma
logo are trademarks of Bloomsbury Publishing Plc

First published in the United Kingdom in 2021.
This edition published 2022

Copyright © Grace Lindsay, 2021

Grace Lindsay has asserted her right under the Copyright, Designs and Patents Act,
1988, to be identified as Author of this work

All rights reserved. No part of this publication may be reproduced or transmitted in any
form or by any means, electronic or mechanical, including photocopying, recording, or
any information storage or retrieval system, without prior permission in writing from
the publishers

Bloomsbury Publishing Plc does not have any control over, or responsibility for, any
third-party websites referred to or in this book. All internet addresses given in this
book were correct at the time of going to press. The author and publisher regret any
inconvenience caused if addresses have changed or sites have ceased to exist, but can
accept no responsibility for any such changes

A catalogue record for this book is available from the British Library

Library of Congress Cataloguing-in-Publication data has been applied for

ISBN: PB: 978-1-4729-6643-8; eBook: 978-1-4729-6645-2

4 6 8 10 9 7 5

Illustrations by Julian Baker

Typeset by Deanta Global Publishing Services, Chennai, India
Printed and bound in Great Britain by CPI Group (UK) Ltd. Croydon, CR0 4YY

Bloomsbury Sigma, Book Sixty-one

To find out more about our authors and books visit www.bloomsbury.com
and sign up for our newsletters

Contents

CHAPTER ONE

Spherical Cows

What mathematics has to offer

The web-weaving spider *Cyclosa octotuberculata* inhabits several locations in and around Japan. About the size of a fingernail and covered in camouflaging specks of black, white and brown, this arachnid is a crafty predator. Sitting at the hub of its expertly built web, it waits to feel vibrations in the web's threads that are caused by struggling prey. As soon as the spider senses this movement, it storms off in the direction of the signal, ready to devour its catch.

Sometimes prey is more commonly found in one location on the web than others. Smart predators know to keep track of these regularities and exploit them. Certain birds, for example, will recall where food has been abundant recently and return to those areas at a later time. *Cyclosa octotuberculata* does something similar – but not identical. Rather than remembering the locations that have fared well – that is, rather than storing these locations in its mind and letting them influence its future attention – the spider literally weaves this information into its web. In particular, it uses its legs to tug on the specific silk threads from which prey has recently been detected, making them tighter. The tightened threads are more sensitive to vibrations, making future prey easier to detect on them.

Making these alterations to its web, *Cyclosa octotuberculata* offloads some of the burden of cognition to its environment. It expels its current knowledge and memory into a compact yet meaningful physical form, making a mark on the world that can guide its future actions. The interacting system of the spider and its web is smarter than the spider could hope to be on its own. This outsourcing of intellect to the environment is known as 'extended cognition'.

Mathematics is a form of extended cognition.

When a scientist, mathematician or engineer writes down an equation, they are expanding their own mental capacity. They are offloading their knowledge of a complicated relationship on to symbols on a page. By writing these symbols down, they leave a trail of their thinking for others and for themselves in the future. Cognitive scientists hypothesise that spiders and other small animals rely on extended cognition because their brains are too limited to do all the complex mental tasks required to thrive in their environment. We are no different. Without tools like mathematics our ability to think and act effectively in the world is severely limited.

Mathematics makes us better in some of the same ways written language does. But mathematics goes beyond everyday language because it is a language that can do real work. The mechanics of mathematics – the rules for rearranging, substituting and expanding its symbols – are not arbitrary. They are a systematic way to export the process of thinking to paper or machines. Alfred Whitehead, a revered twentieth-century mathematician whose work we will encounter in Chapter 3, has been paraphrased as saying: 'The ultimate goal of mathematics is to eliminate any need for intelligent thought.'

Given this useful feature of mathematics, some scientific subjects – physics chief among them – have developed an ethos centred on rigorous quantitative thinking. Scientists in these fields have capitalised on the power of mathematics for centuries. They know that mathematics is the only language precise and efficient enough to describe the natural world. They know that the specialised notation of equations expertly compresses information, making an equation like a picture: it can be worth a thousand words. They also know that mathematics keeps scientists honest. When communicating through the formalism of mathematics, assumptions are laid bare and ambiguities have nowhere to hide. In this way, equations force clear and coherent thinking. As Bertrand Russell (a colleague of Whitehead whom we will also meet in Chapter 3) wrote: 'Everything is vague to a degree you do not realise till you have tried to make it precise.'

The final lesson that quantitative scientists have learnt is that the beauty of mathematics lies in its ability to be both specific and universal. An equation can capture exactly how the pendulum in the barometrical clock on the Ministers' Landing at Buckingham Palace will swing; the very same equation describes the electrical circuits responsible for broadcasting radio stations around the world. When an analogy exists between underlying mechanisms, equations serve as the embodiment of that analogy. As an invisible thread tying together disparate topics, mathematics is a means by which advances in one field can have surprising and disproportionate impacts on other, far-flung areas.

Biology – including the study of the brain – has been slower to embrace mathematics than some other fields. A certain portion of biologists, for reasons good and bad, have historically eyed mathematics with some scepticism. In their opinion, mathematics is both too complex and too simple to be of much use.

Some biologists find mathematics too complex because – trained as they are in the practical work of performing lab experiments and not in the abstract details of mathematical notion – they see lengthy equations as meaningless scribble on the page. Without seeing the function in the symbols, they'd rather do without them. As biologist Yuri Lazebnik wrote in a 2002 plea for more mathematics in his field: 'In biology, we use several arguments to convince ourselves that problems that require calculus can be solved with arithmetic if one tries hard enough and does another series of experiments.'

Yet mathematics is also considered too simple to capture the overwhelming richness of biological phenomena. An old joke among physicists highlights the sometimes absurd level of simplification that mathematical approaches can require. The joke starts with a dairy farmer struggling with milk production. After trying everything he could think to get his beloved cows to produce more, he decides to ask the physicist at the local university for help. The physicist listens carefully to the problem and goes back to his office to think. After some consideration, he comes back to the farmer and says: 'I found a solution. First, we must assume a spherical cow in a vacuum … '

Simplifying a problem is what opens it up to mathematical analysis, so inevitably some biological details get lost in translation from the real world to the equations. As a result, those who use mathematics are frequently disparaged as being too disinterested in those details. In his 1897 book *Advice for a Young Investigator*, Santiago Ramón y Cajal (the father of modern neuroscience whose work is discussed in Chapter 9) wrote about these reality-avoiding theorists in a chapter entitled 'Diseases of the Will'. He identified their symptoms as 'a facility for exposition, a creative and restless imagination, an aversion to the laboratory, and an indomitable dislike for concrete science and seemingly unimportant data'. Cajal also lamented the theorist's preference for beauty over facts. Biologists study living things that are abundant with specific traits and nuanced exceptions to any rule. Mathematicians – driven by simplicity, elegance and the need to make things manageable – squash that abundance when they put it into equations.

Oversimplification and an obsession with aesthetics are legitimate pitfalls to avoid when applying mathematics to the real world. Yet, at the same time, the richness and complexity of biology is exactly why it needs mathematics.

Consider a simple biological question. There are two types of animals in a forest: rabbits and foxes. Foxes eat rabbits, rabbits eat grass. If the forest starts off with a certain number of foxes and a certain number of rabbits, what will happen to these two populations?

Perhaps the foxes ferociously gobble up the rabbits, bringing them to extinction. But then the foxes, having exhausted their food source, will starve and die off themselves. This leaves us with a rather empty forest. On the other hand, maybe the fox population isn't so ravenous. Perhaps they reduce the rabbit population to almost zero but not quite. The fox population still plummets as each individual struggles to find the remaining rabbits. But then – with most of the foxes gone – the rabbit population can rebound. Of course, now the food for the foxes is abundant again and, if enough of their population remains, they too can resurge.

When it comes to knowing the outcome for the forest, there is a clear limitation to relying on intuition. Trying to 'think through' this scenario, as simple as it is, using just words and stories is insufficient. To make progress, we must define our terms precisely and state their relationships exactly – and that means we're doing mathematics.

In fact, the mathematical model of predator–prey interactions that can help us here is known as the Lotka–Volterra model and it was developed in the 1920s. The Lotka-Volterra model consists of two equations: one that describes the growth of the prey population in terms of the numbers of prey and predators, and another that describes the growth of the predator population in terms of the numbers of predators and prey. Using dynamical systems theory – a set of mathematical tools initially forged to describe the interactions of celestial bodies – these equations can tell us whether the foxes will eventually die off, or the rabbits will, or if they'll carry on in this dance together forever. In this way, the use of mathematics makes us better at understanding biology. Without it, we are sadly limited by our own innate cognitive talents. As Lazebnik wrote: 'Understanding [a complex] system without formal analytical tools requires geniuses, who are so rare even outside biology.'

To look at a bit of biology and see how it can be reduced to variables and equations requires creativity, expertise and discernment. The scientist must see through the messy details of the real world and find the bare-bones structure that underlies it. Each component of their model must be defined appropriately and exactly. Once a structure is found and an equation written, however, the fruits of this discipline are manifest. Mathematical models are a way to describe a theory about how a biological system works precisely enough to communicate it to others. If this theory is a good one, the model can also be used to predict the outcomes of future experiments and to synthesise results from the past. And by running these equations on a computer, models provide a 'virtual laboratory', a way to quickly and easily plug in different values to see how different scenarios may turn out and even perform 'experiments' not yet feasible in the physical world. By working through scenarios and hypotheses digitally this way, models help scientists determine what parts of a system are important to its function and, importantly, which are not.

Such integral work could hardly be carried out using simple stories unaccompanied by mathematics. As Larry Abbott, a prominent theoretical neuroscientist and co-author[*] of one of the most widely used textbooks on the subject, explained in a 2008 article:

Equations force a model to be precise, complete and self-consistent, and they allow its full implications to be worked out. It is not difficult to find word models in the conclusions sections of older

---

[*] Along with Peter Dayan, whom we will meet in Chapter 11.

neuroscience papers that sound reasonable but, when expressed as
mathematical models, turn out to be inconsistent and unworkable.
Mathematical formulation of a model forces it to be self-consistent
and, although self-consistency is not necessarily truth, self-
inconsistency is certainly falsehood.

The brain – composed of (in the case of humans) some 100 billion neurons, each their own bubbling factory of chemicals and electricity, all interacting in a jumble of ways with their neighbours both near and far – is a prime example of a biological object too complex to be understood without mathematics. The brain is the seat of cognition and consciousness. It is responsible for how we feel, how we think, how we move, who we are. It is where days are planned, memories are stored, passions are felt, choices are made, words are read. It is the inspiration for artificial intelligence and the source of mental illness. To understand how all this can be accomplished by a single complex of cells, interfacing with a body and the world, demands mathematical modelling at multiple levels.

Despite the hesitancy felt by some biologists, mathematical models can be found hidden in all corners of the history of neuroscience. And while it was traditionally the domain of adventurous physicists or wandering mathematicians, today 'theoretical' or 'computational' neuroscience is a fully developed subdivision of the neuroscience enterprise with dedicated journals, conferences, textbooks and funding sources. The mathematical mindset is influencing the whole of the study of the brain. As Abbott wrote: 'Biology used to be a refuge for students fleeing mathematics, but now many life sciences students have a solid knowledge of basic mathematics and computer programming, and those that don't at least feel guilty about it.'*

Yet the biologist's apprehension around mathematical models should not be entirely dismissed. 'All models are wrong,' starts the popular phrase by statistician George Box. Indeed, all models *are* wrong, because all models ignore some details. All models are also wrong because they represent only a biased view of the processes they claim to capture. And all models are wrong because they favour simplicity over absolute accuracy. All models are wrong the same way all poems are wrong; they capture an essence, if not a perfect literal truth. 'All models are wrong but some are useful,' says Box. If the farmer in the old joke reminded the physicist that cows are not, in fact, spherical, the physicist's response would be, 'Who cares?', or more accurately, 'Do we need to care?'. Detail for detail's sake is not a virtue. A map the size of the city has no good use. The art of mathematical modelling is in deciding which details matter and steadfastly ignoring those that do not.

This book charts the influence of mathematical thinking – borrowed from physics, engineering, statistics and computer science – on the study of the brain. Each chapter tells, for a different topic in neuroscience, the story of the biology, the mathematics and the interplay between the two. No special knowledge of mathematics is assumed

---

* This guilt may not be entirely new. Charles Darwin, certainly a successful biologist, wrote in an 1887 autobiography: 'I have deeply regretted that I did not proceed far enough at least to understand something of the great leading principles of mathematics, for men thus endowed seem to have an extra sense.'

on the part of the reader; the ideas behind the equations
will be explained.* No single theory of the brain is being
proposed; different models solve different problems and
offer complementary approaches to understanding.

The chapters are ordered from the low to the high
level: from the physics of single cells up to the mathematics
of behaviour. The stories in these chapters include the
struggles encountered in unifying mathematics and
biology, and the scientists who did the struggling. They
show that sometimes experiments inform models and
sometimes models inform experiments. They also show
that a model can be anything from a few equations
confined to a page to countless lines of code run on
supercomputers. In this way, the book is a tapestry of the
many forms mathematical models of the brain can take.
Yet while the topics and models covered are diverse,
common themes do reappear throughout the pages.

Of course, everything in this book may be wrong. It
may be wrong because it is science and our understanding
of the world is ever-evolving. It may be wrong because
it is history and there is always more than one way to tell
a story. And, most importantly, it *is* wrong because it is
mathematics. Mathematical models of the mind do not
make for perfect replicas of the brain, nor should we
strive for them to be. Yet in the study of the most
complex object in the known universe, mathematical
models are not just useful but absolutely essential. The
brain will not be understood through words alone.

---

* However, for the mathematically inclined, an appendix elaborating
on one of the main equations per chapter is provided at the end of
the book.

CHAPTER TWO

How Neurons Get Their Spike

Leaky integrate-and-fire and Hodgkin-Huxley neurons

'The laws of action of the nervous principle are totally different from those of electricity,' concluded Johannes Müller more than 600 pages into his 1840 textbook *Handbuch der Physiologie des Menschen*. 'To speak, therefore, of an electric current in the nerves, is to use quite as symbolical an expression as if we compared the action of the nervous principle with light or magnetism.'

Müller's book – a wide-ranging tour through the new and uncertain terrain of the field of physiology – was widely read. Its publication (especially its near-immediate translation into English under the title *Elements of Physiology*) cemented Müller's reputation as a trusted teacher and scientist.

Müller was a professor at Humboldt University of Berlin from 1833 until his death 25 years later. He had a broad interest in biology and strong intellectual views. He was a believer in vitalism, the idea that life relied on a *Lebenskraft*, or vital organising force, that went beyond mere chemical and physical interactions. This philosophy could be found streaking through his physiology. In his book, he claims not only that the activity of nerves is not electric in nature, but that it may ultimately be 'imponderable', the question of its essence 'not capable of solution by physiological facts'.

Müller, however, was wrong. Over the course of the following century, the spirit that animated the nerves would prove wholly reducible to the simple movement of charged particles. Electricity is indeed the ink in which the neural code is written. The nervous principle was perfectly ponderable after all.

More than merely striking down Müller's vitalism, the identification of this 'bio-electricity' in the nervous system provided an opportunity. By forging a path between the two rapidly developing studies of electricity and physiology, it allowed for the tools of the former to be applied to the problems of the latter. Specifically, equations – whittled down by countless experiments to capture the essential behaviours of wires, batteries and circuits – now offered a language in which to describe the nervous system. The two fields would come to share symbols, but their relationship was far more than the merely symbolic one Müller claimed. The proper study of the nervous system depended on collaboration with the study of electricity. The seeds of this collaboration, planted in the nineteenth century, would come to sprout in the twentieth and bloom in the twenty-first.

★ ★ ★

Walk into the home of an educated member of upper-class society in late eighteenth-century Europe and you may find, among shelves of other scientific tools and curiosities, a Leyden jar. Leyden jars, named after the Dutch town that was home to of one of their inventors, are glass jars like most others. However instead of storing jam or pickled vegetables, Leyden jars store charge.

Developed in the mid-eighteenth century, these devices marked a turning point in the study of electricity. As a literal form of lightning in a bottle, they let scientists and non-scientists alike control and transmit electricity for the first time – sometimes doling out shocks large enough to cause nosebleeds or unconsciousness.

While its power may be large, the Leyden jar's design is simple (see Figure 1). The bottom portion of the inside of the jar is covered in a metal foil, as is the same region on the outside. This creates a sandwich of glass in between the two layers of metal. Through a chain or rod inserted at the top of the jar, the internal foil gets pumped full of charged particles. Particles of opposite charge are attracted to each other, so if the particles going into the jar are positively charged, for example, then negatively charged ones will start to accrue on the outside. The particles can never reach each other, however, because the glass of the jar keeps them apart. Like two neighbourhood dogs separated by a fence, they can only line up on either side of the glass, desperately wishing to be closer.

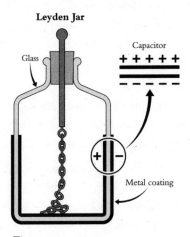

Figure 1

We would now call a device that stores charge like the Leyden jar a 'capacitor'. The disparity in charge on either side of the glass creates a difference in potential energy known as voltage. Over time, as more and more charge is added to the jar, this voltage increases. If the glass barrier disappeared – or another path were provided for these particles to reach each other – that potential energy would turn into kinetic energy as the particles moved towards their counterparts. The higher the voltage was across the capacitor, the stronger this movement of charge – or current – would be. This is exactly how so many scientists and tinkerers ended up shocking themselves. By creating a link between the inside and outside of the jar with their hand, they opened a route for the flow of charged particles right through their body.

Luigi Galvani was an Italian scientist born in 1737. Strongly religious throughout his life, he considered joining the church before eventually studying medicine at the University of Bologna. There he was educated not just in techniques of surgery and anatomy, but also in the fashionable topic of electricity. The laboratory he kept in his home – where he worked closely with his wife Lucia, the daughter of one of his professors – contained devices for exploring both the biological and the electric: scalpels and microscopes, along with electrostatic machines and, of course, Leyden jars. For his medical experiments, Galvani – like students of biology for centuries before and after him – focused on frogs. The muscles in a frog's legs can keep working after death, a desirable feature when trying to simultaneously understand the workings of an animal and dissect it.

It was a result of his lab's diversity – and potentially disorganisation – that landed Galvani in the pages of science textbooks. As the story goes, someone in the lab (possibly Lucia) touched a metal scalpel to the nerve of a dead frog's leg at the exact moment that an errant spark from an electrical device caused the scalpel to carry charge. The leg muscles of the frog immediately contracted, an observation Galvani decided to enthusiastically pursue. In his 1791 book he describes many different preparations for his follow-up experiments on 'animal electricity', including comparing the efficacy of different types of metal in eliciting contractions and how he connected a wire to a frog's nerve during a thunderstorm. He watched its legs contract with each lightning flash.

There had always been some hints that life was making use of electricity. Ibn Rushd, a twelfth-century Muslim philosopher, anticipated several scientific findings when he noted that the ability of an electric fish to numb the fishermen in its waters may stem from the same force that pulls iron to a lodestone. And in the years before Galvani's discovery, physicians were already exploring the application of electric currents to the body as a cure for everything from deafness to paralysis. But Galvani's varied set of experiments took the study of bio-electricity beyond mere speculation and guesswork. He gathered the evidence to show that animal movement follows from the movement of electricity in the animal. He thus concluded that electricity was a force intrinsic to animals, a kind of fluid that flowed through their bodies as commonly as blood.

In line with the spirit of amateur science at the time, upon hearing news of Galvani's work many people set out to replicate it. Putting their personal Leyden jars in contact with any frog they could capture, curious laymen saw the same contractions and convulsions as Galvani did. So broad was the impact of Galvani's work – and along with it the idea of electrical animation – it made its way into the mind of English writer Mary Shelley, forming part of the inspiration for her novel *Frankenstein*.

A healthy dose of scientific scepticism, however, meant that not all of Galvani's academic peers were so enthusiastically accepting of his claims. Alessandro Volta – an Italian physicist after whom 'voltage' was named – acknowledged that electricity could indeed cause muscle contractions in animals. But he denied that this means animals *normally* use electricity to move. Volta didn't see in Galvani's experiments any evidence that animals were producing their own electricity. In fact, he found that contact between two different metals could create many, nearly imperceptible, electric forces and therefore any test of animal electricity using metals in contact could be contaminated by externally generated electricity. As Volta wrote in an 1800 publication: 'I found myself obliged to combat the pretended animal electricity of Galvani and to declare it an external electricity moved by the mutual contact of metals of different kinds'.*

Unfortunately for Galvani, Volta was a younger man, more willing to engage in public debate and on his way up in the field. He was a formidable scientific opponent. The power of Volta's personality meant Galvani's ideas, though correct in many ways, would be eclipsed for decades.

---

* In the course of proving that contact between dissimilar metals generates electricity, Volta ended up inventing the battery.

Müller's textbook came nearly 10 years after Volta's death, but his objection to animal electricity followed similar lines. He simply didn't believe electricity was the substance of nervous transmission and the weight of the evidence at the time couldn't sway him. In addition to his vitalist tendencies, this stubbornness was perhaps due to Müller's preference for observation over intervention. No matter how many examples of animals responding to externally applied electricity amassed over the years, it would never equal a direct observation of an animal generating its own electricity. 'Observation is simple, indefatigable, industrious, upright, without any preconceived opinion,' said Müller in his inaugural lecture at the University of Bonn. 'Experiment is artificial, impatient, busy, digressive, passionate, unreliable.' At the time, however, observation was impossible. No tool was powerful enough to pick up on the faint electrical signals carried by nerves in their natural state.

That changed in 1847 when Emil du Bois-Reymond – one of Müller's own students – fashioned a very sensitive galvanometer,* a device that measures current through its interaction with a magnetic field. His experiments were an attempt to replicate in nerves what Italian physicist Carlo Matteucci had recently observed in muscles. Using a galvanometer, Matteucci detected a small change in current coming from muscles after forcing them to contract. Searching for this signal in a nerve, however, demanded a stronger magnetic field to pick up the weaker current. In addition to designing proper insulation to prevent any interference from outside electricity,

---

* Named, of course, after our man Galvani.

du Bois-Reymond had to coil more than a mile of wire by hand (producing more than eight times the coils of Matteucci) to get a magnetic field strong enough for his purposes. His handiwork paid off. With his galvanometer measuring its response, du Bois-Reymond stimulated a nerve in various ways – including electrically or with chemicals like strychnine – and monitored the galvanometer's reading of how the nerve responded. Each time, he saw the galvanometer's needle shoot up. Electricity had been spotted at work in the nervous system.

Du Bois-Reymond was a showman as much as he was a scientist and he lamented the dry presentation styles of his fellow scientists. To spread the fruits of his labour, he built several public-ready demonstrations of bio-electricity, including a set-up where he could make a needle move by squeezing his arm in a jar of salt water. All this helped ensure that his findings would be noticed and that du Bois-Reymond would be fondly regarded by the minds of his time. As he said: 'Popularisers of science persist in the public mind as memorial stones of human progress long after the waves of oblivion have surged over the originators of the soundest research.'

Luckily his research was sound as well. Particularly, the follow-up work du Bois-Reymond carried out with his student Julius Bernstein would seal the fate of the theory of nervous electricity. Du Bois-Reymond's original experiment had succeeded in showing a signature of current change in an activated nerve. But Bernstein, through clever and careful experimental design, was able to both amplify the strength of the signal and record it at a finer timescale – creating the first true observation of the elusive nervous signal.

Bernstein's experiments worked by first isolating a nerve and placing it on to his device. The nerve was then electrically stimulated at one end and Bernstein would look for the presence of any electrical activity some distance away. By recording with a precision of up to one-third of one-thousandth of a second, he saw how the nerve current changed characteristically over time after each stimulation. Depending on how far his recording site was from the stimulation site, there may be a brief pause as the electric event travelled down the nerve to reach the galvanometer. Once it got to where he was recording, however, he always saw the current rapidly decrease and then more slowly recover to its normal value.

Bernstein's result, published in the inaugural issue of the *European Journal of Physiology* in 1868, was the first known recording of what is now referred to as an 'action potential'. An action potential is defined as a characteristic pattern of changes in the electrical properties of a cell. Neurons have action potentials. Certain other excitable cells, like those in the muscles or heart, do as well.

This electrical disturbance travels across a cell's membrane like a wave. In this way, action potentials help a cell carry a signal from one end of itself to the other. In the heart, for example, the ripple of an action potential helps coordinate a cell's contraction. Action potentials are also a way for a cell to say something to other cells. In a neuron, when an action potential reaches the knobbly end of an outgrowth known as an axon, it pushes out neurotransmitters. These chemicals can reach other cells and trigger action potentials in them as well. In the case of the familiar frog nerve, the action potentials travelling down the leg lead to the release of neurotransmitters on to the leg muscle. Action potentials in the muscle then cause it to twitch.

Bernstein's work was the first word in a long story about the action potential. Now recognised as the core unit of communication in the nervous system, the action potential forms the basis of modern neuroscience. This quick blip of electrical activity connects the brain to the body, the body to the brain and links all the neurons of the brain in between.

With his glimpsing of current changes coming from the nerve, du Bois-Reymond wrote: 'If I do not greatly deceive myself, I have succeeded in realising [...] the hundred years' dream of physicists and physiologists, to wit, the identity of the nervous principle with electricity.' The nervous principle was indeed identified in the action potential. Yet du Bois-Reymond had committed himself to a 'mathematico-physical method' of explaining biology, and while he had established the physical, he had not quite solved the mathematical. With a growing sense among scientists that proper science involved quantification, the job of describing the physical properties of the nervous principle was far from done. Indeed, it would take roughly another hundred years to capture the essence of the nervous principle in equations.

★ ★ ★

In contrast to the experience of Johannes Müller, when Georg Ohm published a book of his scientific findings he lost his job.

Ohm was born in 1789, the son of a locksmith. He studied for only a short time at the university of his hometown, Erlangen in Germany, and then spent years teaching mathematics and physics in various cities. Eventually, with an aim of becoming an academic, he started performing his own small experiments, particularly around the topic of electricity. For one test, he cut wires of various lengths out of different metals. He then applied voltage across the two ends of the wire and measured how much current flowed between them. Through this he was able to deduce a mathematical relationship between the length of a wire and its current: the longer the wire, the lower the current.

By 1827, Ohm had collected this and other equations of electricity into his book, *The Galvanic Circuit Investigated Mathematically*. Quite contrary to its modern form, the study of electricity in the time of Ohm wasn't a very mathematical discipline and Ohm's peers didn't like his attempts to make it one. One reviewer went so far as to say: 'He who looks on the world with the eye of reverence must turn aside from this book as the result of an incurable delusion, whose sole effort is to detract from the dignity of nature.' Having taken time away from his job to write the book in the hope it would land him a promotion, Ohm, with the book's failure, ended up resigning instead.

Ohm, however, was right. The key relationship he observed – that the current that runs through a wire is equal to the voltage across it divided by the wire's resistance – is a cornerstone of modern electrical engineering taught to first-year physics students worldwide. This is now known as Ohm's law and the standard unit of resistance is the 'ohm'. Ohm wouldn't know the full impact of his work in his lifetime, but he did eventually get some recognition. At the age of 63 he was finally appointed a professor of experimental physics at the University of Munich, two years before he died.

Resistance, as the name suggests, is a measure of opposition. It's a description of just how much a material impedes the course of current. Most materials have some amount of resistance, but, as Ohm noted, the physical properties of a material determine just how resistant it is. Longer wires have higher resistance; thicker ones have lower. Just as the narrowing of an hourglass slows the flow of sand, wires with higher resistance hinder the flow of charged particles.

Louis Lapicque knew of Ohm's law. Born in France in 1866, shortly after the first recording of an action potential, Lapicque completed his doctorate at the Paris Medical School. He wrote his thesis on liver function and iron metabolism. Though his studies were scientific, his interests ranged more broadly from history to politics to sailing; he sometimes even took his boat to conferences across the English Channel.

It was around the start of the twentieth century that Lapique started studying the nerve impulse. It would be the beginning of a decades-long project with his student-turned-wife-and-colleague Marcelle de Heredia, which centred on the concept of time in nerves. One of their earliest questions was: how long does it take to activate a nerve? It was well established by then that applying voltage across a nerve[*] caused a response – measured either as an action potential observed directly in the

---

[*] Applying voltage was an easier way to control the flow of charge than injecting current directly.

nerve or as a muscle twitch that resulted from it. It was also clear that the amount of voltage applied mattered: higher voltage and the nerve would respond quicker, lower and it would respond slower. But what was the exact *mathematical* relationship between the stimulation applied and the time it took to get a response?

This may sound like something of a small research question, a curiosity not of much consequence, but it was Lapicque's approach to it that mattered. Because a proper physiologist also needed to be an engineer – designing and building all manner of electrical devices for stimulating and recording from nerve fibres – Lapicque knew the rules of electricity. He knew about capacitors, resistance, voltage and Ohm's law. And it was with this knowledge that he composed a mathematical concept of the nerve that would answer his question – and many more to come.

Understanding of the membranes that enclose cells had grown in the decades before Lapicque's work. It was becoming clear that these bundles of biological molecules worked a little bit like a brick wall: they didn't let much through. Some of the particles they were capable of keeping apart included ions – atoms of different elements like chloride, sodium or potassium that carry a positive or negative charge. So, just as charged particles could build up on either side of the glass in a Leyden jar, so too could they accrue on the inside and outside of a cell. As Lapicque wrote in his 1907 paper: 'These ideas lead, when treated in the simplest possible way, to already established equations for the polarisation of metal electrodes.'

He thus came to describe the nerve in terms of an 'equivalent circuit'. (see Figure 2) That is, he assumed

that different parts of the nerve acted like the different
components of an electrical circuit. The first equivalence
was made between the cell membrane and a capacitor, as
the membrane could store charge in just the same way.
But it was clear that these membranes weren't acting as
perfect capacitors; they couldn't keep all of the charge
apart. Instead, some amount of current seemed to flow
between the inside and outside of the cell, allowing it to
discharge slightly. A wire with some resistance could play
this role. So Lapicque added a resistor to his circuit
model of the nerve in parallel with the capacitor. This
way, when current is injected into the circuit, some of
that charge goes to the capacitor and some goes through
the resistor. Trying to create a charge difference across
the cell is therefore like pouring water into an imperfect
bucket; much of it would stay in the bucket, but some
would leak away.

This analogy between a cell and a circuit made it
possible for Lapicque to write down an equation. The
equation described how the voltage across the cell

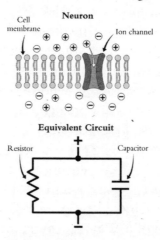

Figure 2

membrane should change over time, based on how much voltage was being applied to it and for how long. With this formalisation in mind, he could calculate when the nerve would respond.

For the data to test his equation on, Lapicque turned to the standard frog-leg experiment: he applied different amounts of voltage to the frog's nerve and recorded the time it took to see a response. Lapicque assumed that when the frog nerve responded it was because the voltage across its membrane had reached a certain threshold. He therefore calculated how long his model would take to reach that threshold for each different voltage applied. Comparing the predictions from his model with the results of his experiments, Lapicque found a good match. He could predict just how long a certain voltage would need to be applied in order to make the nerve respond.

Lapicque wasn't the first to write down such an equation. A previous scientist, Georges Weiss, offered a guess as to how to describe this relationship between voltage and time. And it was a relatively good guess too; it deviated from Lapicque's predictions only somewhat, for example, in the case of voltages applied for a long time. But just as the smallest clue at a crime scene can change the narrative of the whole event, this slight difference between the predictions of Lapicque's equation and what came before actually signalled a deep difference in understanding.

Unlike Lapicque's, Weiss' equation wasn't inspired by the mechanics of a cell nor was it meant to be interpreted as an equivalent circuit. It was more a description of the data than a model of it. Whereas a descriptive equation is like a cartoon animation of an event – capturing its appearance but without any depth – a model is a re-enactment. A mathematical model of a nerve impulse thus needs to have the same moving parts as the nerve itself. Each variable should be mappable to a real physical entity and their interactions should mirror the real world as well. That is just what Lapicque's equivalent circuit provided: an equation where the terms were interpretable.

Others before Lapicque had seen the similarity between the electrical tools used to study the nerve and the nerve itself. Lapicque was building heavily on the work of Walther Nernst, who noticed that the membrane's ability to separate ions could underlie the action potential. Another student of du Bois-Reymond, Ludimar Hermann, had spoken of the nerve in terms of capacitors and resistors. And even Galvani himself had a vision of a nerve that worked similarly to his Leyden jar. But with his explicit equivalent circuit and quantitative fit to data, Lapicque went a step further in making an argument for the nerve as a precise electrical device. As he wrote: 'The physical interpretation that I reach today gives a precise meaning to several important previously known facts on excitability ... It seems to me a reason to consider it a step in the direction of realism.'

Due to their limited equipment, most neuroscientists of Lapicque's time were recording from whole nerves. Nerves are bundles of many axons – the fibres through which individual neurons send their signals to other cells. Recording from many axons at once makes it easier to pick up the current changes they produce, but harder to see the detailed shape of those changes. Sticking an electrode into a single neuron, however, makes it possible to record the voltage across its membrane directly. Once the technology to observe individual neurons became available in the early twentieth century, the action potential came into much clearer view.

One defining feature of the action potential noticed by English physiologist Edgar Adrian in the 1920s is the 'all-or-nothing' principle.* The all-or-nothing principle says that a neuron either emits an action potential or it doesn't – nothing in between. In other words, any time a neuron gets enough input, the voltage across its membrane changes – and it changes in exactly the same way. So, just as a goal in hockey counts the same no matter how hard the puck is driven into the net, strongly stimulating a neuron doesn't make its action potential any bigger or better. All stronger stimulation can do is make the neuron emit *more* of the exact same action potentials. In this way, the nervous system cares more about quantity than quality.

The all-or-nothing nature of a neuron aligns with Lapicque's intuition about a threshold. He knew that the voltage across the membrane needed to reach a certain value in order to see a response from the nerve. But once it got there, a response was a response.

By the 1960s, the all-or-nothing principle was combined with Lapicque's equation into a mathematical model known as a 'leaky integrate-and-fire neuron': 'leaky' because the presence of a resistor means some of the current leaks away; 'integrate' because the capacitor integrates the rest of it and stores it as charge; and 'fire'

---

* More on Adrian, and what his discovery meant about how neurons represent information, in Chapter 7.

because when the voltage across the capacitor reaches the threshold the neuron 'fires', or emits an action potential. After each 'firing' the voltage is reset to its baseline, only to reach threshold again if more input is given to the neuron.

While the model is simple, it can replicate features of how real neurons fire: for example, with strong and constant input the model neuron will fire action potentials repeatedly, with only a small delay between each one; if the input is low enough, however, it can remain on indefinitely without ever causing a single action potential.

These model neurons can also be made to form connections – strung together such that the firing of one generates input to another. This provides modellers with a broader power: to replicate, explore and understand the behaviour of not just individual neurons but whole networks of them.

Since their inception, such models have been used to understand countless aspects of the brain, including disease. Parkinson's disease is a disorder that impacts the firing of neurons in the basal ganglia. Located deep in the brain, the basal ganglia are composed of a variety of regions with elaborate Latin names. When the input to one region – the striatum – is perturbed by Parkinson's disease, it knocks the rest of the basal ganglia off balance. As a result of changes to the striatum, the subthalamic nucleus (another region of the basal ganglia) starts to fire more, which causes neurons in the globus pallidus external (yet another basal ganglia region) to fire. But those neurons send connections *back* to the subthalamic nucleus that *prevent* those neurons from firing more – which also, in turn, shuts down the globus pallidus external itself. The result of this complicated web of connections is oscillations: neurons in this network fire more, then less, then more again. These rhythms appear to be related to the movement problems Parkinson's patients have – tremors, slowed movements and rigidity.

In 2011, researchers at the University of Freiburg built a computer model of these brain regions made up of 3,000 leaky integrate-and-fire model neurons. In the model, disturbing the cells that represented the striatum created the same problematic waves of activity seen in the subthalamic nuclei of Parkinson's patients. With the model exhibiting signs of the disease, it could also be used to explore ways to treat it. For example, injecting pulses of input into the model's subthalamic nucleus broke these waves down and restored normal activity. But the pulses had to be at just the right pace – too slow and the offending oscillations got worse, not better. Deep brain stimulation – a procedure where pulses of electrical activity are injected into the subthalamic nucleus of Parkinson's patients – is known to help alleviate tremors. Doctors using this treatment also know that the rate of pulses has to be high, around 100 times per second. This model gives a hint as to why high rates of stimulation work better than lower ones. In this way, modelling the brain as a series of interconnected circuits illuminates how the application of electricity can fix its firing.

Lapicque's original interest was in the timing of neural firing. By piecing together the right components of an electrical circuit he captured the timing of action potentials correctly, but the creation of this circuit stand-in for a neuron did more than that. It formed a solid foundation on which to build towering networks of thousands of interconnecting cells. Now computers across the world churn through the equations of these faux neurons, simulating how real neurons integrate and fire in both health and disease.

★ ★ ★

In the summer of 1939, Alan Hodgkin set out on a small fishing boat off the southern coast of England. His goal was to catch some squid, but mostly what he got was seasick.

At the time, Hodgkin, a research fellow at Cambridge University, had only just arrived at the Marine Biological Association in Plymouth ready to embark on a new project studying the electrical properties of the squid giant axon. In particular, he wished to know how the action potential got its characteristic up–down shape (frequently referred to as a 'spike'[*]). A few weeks later he was joined by an equally green student collaborator, Andrew Huxley. Luckily, the men eventually figured out when and where their subject matter could be found in the sea.

Though Huxley was a student of Hodgkin's, the men were only four years apart in age. Hodgkin looked the role of a proper English gentleman: long face, sharp eyes, his hair neatly parted and swept to the side. Huxley was a bit more boyish with round cheeks and heavy eyebrows. Both men had skill in biology and physics, though each came to this pairing from the opposite side.

Hodgkin primarily studied biology, but in his last term he was encouraged by a zoology professor to learn

---

[*] Spike, firing, activity, action potential – the sacred emission of a neuron goes by many names.

as much mathematics and physics as he could. Hodgkin obliged, spending hours with textbooks on differential equations. Huxley was long interested in mechanics and engineering, but switched to a more biological track after a friend told him physiology classes would teach more lively and controversial topics. Huxley may have also been drawn to these subjects by the influence of his grandfather. Biologist Thomas Henry Huxley – known as 'Darwin's bulldog' for his vociferous defence of Darwin's theory of evolution – described physiology as 'the mechanical engineering of living machines'.

Lapicque's model predicted when a cell would fire, but it still didn't explain exactly what an action potential was. At the time of Hodgkin's boat trip, the going theory of what happens when a neuron emits an action potential was still the one put forward by the original action potential observer himself, Julius Bernstein. It said that, at the time of this electrical event, the cell membrane temporarily breaks down. It therefore lets ions of all different kinds flow through, erasing the charge difference that is normally across it and creating the small current Bernstein saw with his galvanometer.

But some of Hodgkin's previous experiments on crabs told him this may not be quite right. He wanted to follow up on this work with the squid because the large size of the axon running along its mantle made precise measurements easier.* Sticking an electrode into this axon, Hodgkin and Huxley recorded the voltage changes

---

* The 'squid giant axon' that Hodgkin and Huxley were studying is a particularly large (about the width of a marker tip) axon of a rather average-sized squid. It is not, as many students of neuroscience initially believe, the axon of a giant squid.

that occurred during an action potential (See Figure 3). What they saw was a clear 'overshoot'. That is, the voltage didn't just go to zero, as would happen with a discharged capacitor, but rather it *reversed*. While a neuron normally has more positive charge on the outside of the cell than on the inside, during the peak of the action potential this pattern gets inverted and the inside becomes more positively charged than the outside. Simply letting more ions diffuse through the membrane wouldn't lead to such a separation. Something more selective was at play.

Only a short time after Hodgkin and Huxley made this discovery, their work was unfortunately interrupted. Hitler invaded Poland. The men needed to abandon the lab and join the war effort. Solving the mystery of the action potential would have to wait.

When Hodgkin and Huxley returned to Plymouth eight years later the lab required a bit of reassembling: the building had been bombed in air raids and their equipment had been passed around to other scientists. But the men, each having picked up some extra quantitative skills as a result of their wartime assignments – Huxley performing data analysis for the

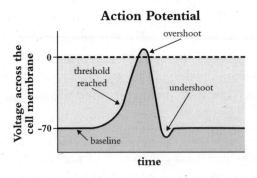

Figure 3

Gunnery Division of the Royal Navy and Hodgkin developing radar systems for the Air Force – were eager to get back to work on the physical mechanisms of the nerve impulse.

For many of the following years, Hodgkin and Huxley (helped by fellow physiologist Bernard Katz) played with ions. By removing a certain type of ion from the neuron's environment, they could determine which parts of the action potential relied on which kinds of charged particles. A neuron kept in a bath with less sodium had less of an overshoot. Add extra potassium to the bath and the undershoot – an effect at the very end of the action potential when the inside of the cell becomes more negative than normal – disappeared. The pair also experimented with a technique that let them directly control the voltage across the cell membrane. Changing this balance of charge created large changes in the flow of ions into and out of the cell. Remove the difference in charge across the membrane and stores of sodium outside the cell suddenly swim inwards; keep the cell in this state a bit longer and potassium ions from inside the cell rush out.

The result of all these manipulations was a model. Specifically, Hodgkin and Huxley condensed their hard-won knowledge of the nuances of neural membranes into the form of an equivalent circuit and with it a corresponding set of equations. This equivalent circuit was more complex than Lapicque's, however. It had more moving parts as it aimed to explain not just *when* an action potential happens but the full shape of the event itself. The main difference, though, came down to resistance.

In addition to the resistor Lapicque put in parallel with the membrane capacitor, Hodgkin and Huxley added two more – one specifically controlling the flow of sodium ions and the other controlling the flow of potassium ions. Such a separation of resistors assumed that different channels in the cell membrane were selectively allowing different ion types to pass. What's more, the strength of these resistors – that is, the extent to which they block the flow of their respective ions – is not a fixed parameter in the model. Instead, they are dependent on the state of the voltage across the capacitor. The cell accomplishes this by opening or closing its ion channels as the voltage across its membrane changes. In this way, the membrane of the cell acts like the bouncer of a club: it assesses the population of particles on either side of itself and uses that to determine which ions get to enter and exit the cell.

Having defined the equations of this circuit, Hodgkin and Huxley wanted to churn through the numbers to see if the voltage across the model's capacitor really would mimic the characteristic whish and whoosh of an action potential. There was a problem, however. Cambridge was home to one of the earliest digital computers, a device that would've greatly sped up Hodgkin and Huxley's calculations, but it was out of service. So, Huxley turned to a Brunsviga – a large, metal calculator powered by a hand-crank. As he sat for days putting in the value of the voltage at one point in time just to calculate what it would be at the next one-ten-thousandth of a second, Huxley actually found the work somewhat suspenseful. As he said in his Nobel lecture: 'It was quite often exciting ... Would the membrane potential get away into a spike, or die in a subthreshold oscillation? Very often my expectations turned out to be wrong, and an important lesson I learnt from these manual computations was the complete inadequacy of one's intuition in trying to deal with a system of this degree of complexity.'

With the calculations complete, Hodgkin and Huxley had a set of artificial action potentials, the behaviour of which formed a near-perfect mirror image of a real neuron's spike.

When injected with current, the Hodgkin-Huxley model cell displays a complex dance of changing voltage and resistances. First, the input fights against the cell's natural state: it adds some positive charge to the largely negative inside of the cell. If this initial disturbance in the membrane's voltage is large enough – that is, if the threshold is met – sodium channels start opening and a glut of positively charged sodium ions flood into the cell. This creates a positive feedback loop: the influx of sodium ions pushes the inside of the cell more positive and the resulting change in voltage lowers the sodium resistance even more. Soon, the difference in charge across the membrane disappears. The inside of the cell is briefly as positive as the outside, and then more so – the 'overshoot'. As this is happening potassium channels are opening, letting positively charged potassium ions fall out of the cell. The sodium and potassium channels work like saloon doors, one letting ions in and the other out, but now the potassium ions are moving quicker. The work of the potassium ions reverses the trend in voltage. As this exodus of potassium again makes the inside of the cell more negative, sodium channels close. The separation of charge across the membrane is being rebuilt. As the voltage nears its original value, positive charge continues to leak out of the still-open potassium channels – the 'undershoot'. Eventually these too close, the voltage recovers and the cell has returned to normal, ready to fire again. The whole event takes less than one-half of one-hundredth of a second.

According to Hodgkin, the pair built this mathematical model because 'at first it might be thought that the response of a nerve to different electrical stimuli is too complicated and varied to be explained by these relatively simple conclusions'. But explain it they did. Like a juggler, the neuron combines simple parts in simple ways to create a splendidly intricate display. The Hodgkin-Huxley model makes clear that the action potential is a delicately controlled explosion occurring a billion times a second in your brain.

The pair published their work – both experimental and computational – in a slew of *Journal of Physiology* papers in 1952. Eleven years later, they were awarded two-thirds of the Nobel Prize for 'their discoveries concerning the ionic mechanisms involved in excitation and inhibition in the peripheral and central portions of the nerve cell membrane'. If doubts remained in the minds of any biologists about whether the nerve impulse was explainable in terms of ions and electricity, the work of Hodgkin and Huxley put those to rest.

★ ★ ★

'The body *and dendrites* of a nerve cell are specialised for the reception and integration of information, which is conveyed as impulses that are fired from other nerve cells along their axons' (emphasis added). With this modest sentence John Eccles, an Australian neurophysiologist and the third awardee alongside Hodgkin and Huxley, began his Nobel lecture. The lecture goes on to describe the intricacies of ion flows that occur when one cell sends information to another.

What the lecture doesn't discuss is dendrites. Dendrites are the wispy tendrils that grow out of a neuron's cell body. These offshoots, like tree roots, branch and stretch and branch again to cover a wide area around the cell. A cell casts its dendritic net out among nearby cells to collect input from them.

Eccles had a complex relationship with dendrites. The type of neuron he studied – found in the spinal cord of cats – had elaborate dendritic branches. They spanned roughly 20 times the size of the cell's body in all directions. Yet Eccles didn't believe this cellular root system was terribly relevant. He conceded that the parts of the dendrites nearest the cell body may have some use: axons from other neurons land on these regions and their inputs will get transported immediately into the cell body where they can contribute to causing an action potential. But, he claimed, those farther out were simply too distant to do much: their signal wouldn't survive the journey to the cell body. Instead, he assumed the cell used these arms to absorb and expel charged particles in order to keep its overall chemical balance intact. In Eccles' eyes, therefore, dendrites were – at most – a wick that carried a flame a short way to the cell body and, at least, a straw slurping up some ions.

Eccles' position on dendrites put him at odds with his student, Wilfrid Rall. Rall earned a degree in physics from Yale University in 1943 but, after his time working on the Manhattan Project, became interested in biology. He moved to New Zealand to work with Eccles on the effects of nerve stimulation in 1949.

Given his background, Rall was quick to turn to mathematical analyses and simulations to understand a system as complex as a biological cell. And he was inspired and galvanised[*] by the work of Hodgkin and Huxley, which he heard about when Hodgkin visited the University of Chicago where Rall was getting his master's degree. With this mathematical model in mind, Rall suspected that dendrites were capable of more than Eccles gave them credit for. After his time in New Zealand, Rall devoted a good portion of his career to proving the power of dendrites – and, in turn, proving the power of mathematical models to anticipate discoveries in biology.

Building on the analogy of the cell as an electrical circuit, Rall modelled the thin cords of dendrites as just what they looked like: cables. The 'cable theory' approach to dendrites treats each section of a dendrite as a very narrow wire, the width of which – just as Ohm discovered – determines its resistance. Stringing these sections together, Rall explored how electrical activity at the far end of a dendrite can make its way to the cell body or vice versa.

Adding more parts to this mathematical model, however, meant more numbers to crunch. The National Institutes of Health (NIH) in Bethesda, Maryland, where Rall worked, didn't have a digital computer suitable for some of his larger simulations. When Rall wanted to run

---

[*] Also named after our man Galvani.

the equations of a model with extensive dendrites, it was the job of Marjory Weiss, a programmer at the NIH, to drive a box of punch cards with the computer's instructions to Washington DC to run them on a computer there. Rall couldn't see the results of his model until she returned the next day.

Through his elaborate mathematical explorations, Rall clearly showed – against the beliefs of Eccles – that a cell body with dendrites can have very different electrical properties than one without. A short description of Rall's calculations, published in 1957, set off a years-long debate between the two men in the form of a volley of publications and presentations.[*] Each pointed to experimental evidence and their own calculations to support their side. But slowly, over time, Eccles' position shifted. By 1966, he had publicly accepted dendrites as a relevant cog in the neural machinery. Rall was right.

Cable theory did more than just expose Eccles' mistake. It also offered a way for Rall to explore in equations the many magical things that dendrites can do before the experimental techniques to do so were available. One important ability Rall identified was detecting order. Rall saw in his simulations that the order in which a dendrite gets input has important consequences for the cell's response. If an input comes to a dendrite's far end first, followed by more inputs closer and closer to the cell body, the cell may fire. If the

---

[*] According to Rall, Eccles even prevented his work from being published. With respect to a 1958 manuscript: 'A negative referee persuaded the editors to reject this manuscript. The fact that this referee was Eccles, was clear from many marginal notes on the returned manuscript.'

pattern is reversed, however, it won't. This is because
inputs coming in far from the cell body take longer to
reach it. So, starting the inputs at the far end means they
all reach the cell body at the same time. This creates a big
change in the membrane's voltage and possibly a spike.
Going the other way, however, inputs come in at different
times; this creates only a middling disturbance in voltage.
In a race where runners start at different times and
locations, the only way to get them to cross the finish
line together is to give the farther ones a head start.

Rall made this prediction in 1964. In 2010, it was
shown to be true in real neurons. To test Rall's hypothesis,
researchers at University College London took a sample
of neurons from rat brains. Placing these neurons in a dish,
they were able to carefully control the release of
neurotransmitters on to specific portions of a dendrite –
portions as little as five microns (or the width of a red
blood cell) apart. When this input went from the end of a
dendrite to its root, the cell spiked 80 per cent of the time.
In the other direction, it responded only half as often.

This work shows how even the smallest bit of biology
has a purpose. That the sections of a dendrite can work
like the keys of a piano – where the same notes can be
played in different ways to different effect – gives
neurons new tricks. Specifically, it imbues neurons with
the ability to identify sequences. There are many
occasions where inputs sweeping across the dendrite
from one direction should be treated differently from
inputs sweeping the other way. For example, neurons in
the retina have this kind of 'direction selectivity'. This
lets them signal which way objects in the visual field
are moving.

In many science classes, students are given small electrical circuit kits to play with. They can use wires of different resistances to connect up capacitors and batteries, maybe making a lightbulb light up or a fan spin. In much the same pick-and-play way, neuroscientists now build models of neurons. With the basic parts list of an electrical circuit, almost any observed property of a neuron's activity can be mimicked. Rall helped add more parts to the kit.

★ ★ ★

If a standard neuron model is a small house built out of the bricks of electrical engineering, then the model constructed by the Blue Brain Project in 2015 is an entire metropolis. Eighty-two scientists across 12 institutions worked together as part of this unprecedented collaboration. Their goal was to replicate a portion of the rat brain about the size of a large grain of sand. They combed through previous studies and spent years performing their own experiments to collect every bit of data they could about the neurons in this region. They identified the ion channels they use, the length of their axons, the shapes of their dendrites, how closely they pack together and how frequently they connect. Through this, they identified 55 standard shapes that neurons can take, 11 different electrical response profiles they can have and a host of different ways they can interact.

They used this data to build a simulation – a simulation that included more than 30,000 highly detailed model neurons forming 36 million connections. The full model required a specially built supercomputer to run through the billions of equations that defined it. Yet all of this complexity still stemmed from the same basic principles of Lapicque, Hodgkin, Huxley and Rall. A lead researcher on the project, Idan Segev, summarised the approach: 'Use Hodgkin-Huxley in an extended way and build a simulation of the way these cells are active, to get the music – the electrical activity – of this network of neurons that should imitate the real biological network that you're trying to understand.'

As the team showed in their publication documenting the work, the model was able to reproduce several features of the real biological network. The simulation showed similar sequences of firing patterns over time, a diversity of responses across cell types and oscillations. More than just replicating results of past experiments, this real-to-life model also makes it possible to explore new experiments quickly and easily. Recreating the biology in a computer makes virtual investigations of this brain region as simple as writing a few lines of code – an approach known as 'in silico' neuroscience.

Running such simulations can only give good predictions if the model underlying them is a reasonable facsimile of biology. Thanks to Lapicque, we know that using the equations of an electrical circuit as a stand-in for a neuron is a solid foundation on which to build models of the brain. It was his analogy that set off the study of the nerve as an electrical device. And the extension of his analogy by countless other scientists – many trained in both physics and physiology – expanded its explanatory power even further. The nervous system – against Müller's intuitions – is brought to life by the flow of electricity and the study of it has been undeniably animated by the study of electricity.

CHAPTER THREE

Learning to Compute

McCulloch–Pitts, the Perceptron and artificial neural networks

Cambridge University mathematician Bertrand Russell spent 10 years at the beginning of the twentieth century toiling towards a monumental goal: to identify the philosophical roots from which all of mathematics stems. Undertaken in collaboration with his former teacher Alfred Whitehead, this ambitious project produced a book, the *Principia Mathematica*, which was delivered to the publishers overdue and over budget. The authors themselves had to chip in towards the publishing costs just to get it done and they didn't see any royalties for 40 years.

But the financial hurdle was perhaps the smallest one to overcome in getting this opus finished. Russell had to fight against his own agitation with the scholarly material. According to his autobiography, he spent days staring at a blank sheet of paper and evenings contemplating jumping in front of a train. Work on the book also coincided with the dissolution of Russell's marriage and a strain on his relationship with Whitehead – who, according to Russell, was fighting his own mental and marital battles at the time. The book was even physically demanding: Russell spent 12 hours a day at a desk writing out the intricate symbolism needed to convey his complex mathematical ideas and when the time came to bring the manuscript to the publisher it was too large for him to carry. Despite it all, Russell and Whitehead eventually completed and published the text they hoped would tame the seemingly wild state of mathematics.

The conceit of the *Principia* was that all of mathematics could be reduced to logic. In other words, Russell and Whitehead believed that a handful of basic statements, known as 'expressions', could be combined in just the right way to generate all the formalisms, claims and findings of mathematicians. These expressions didn't stem from any observations of the real world. Rather, they were meant to be universal. For example, the expression: if X is true, then the statement 'X is true or Y is true' is also true. Such expressions are made up of propositions: fundamental units of logic that can be either true or false, written as letters like X or Y. These propositions are strung together with 'Boolean' operators* such as 'and', 'or' and 'not'.

In the first volume of the *Principia*, Russell and Whitehead provided fewer than two dozen of these abstract expressions. From these humble seeds, they built mathematics. They were even able to triumphantly conclude – after scores of symbol-filled pages – that $1+1=2$.

Russell and Whitehead's demonstration that the full grandeur of mathematics could be captured with the simple

---

* Named after the English mathematician George Boole. While they used his ideas, Russell and Whitehead didn't actually use the term 'Boolean', as it wasn't coined until 1913.

rules of logic* had immense philosophical implications as it provided proof of the power of logic. What's more, it meant that a subsequent finding, made by a different pair of men some 30 years later, would have immense implications of its own. This finding said that neurons, simply via the nature of their anatomy and physiology, were implementing the rules of logic. It revolutionised the study of the brain and of intelligence itself.

★ ★ ★

When Detroit native Walter Pitts was just 12 years old, he was invited by Russell to join him as a graduate student at Cambridge University. The young boy had, the story goes, encountered a copy of the *Principia* after running into a library to avoid bullies. As he read, Pitts found what he believed to be errors in the work. So, he sent his notes on the subject off to Russell, who, presumably not knowing the boy's age, then offered him the position. Pitts didn't accept it. But a few years later, when Russell was visiting the University of Chicago, Pitts went to sit in on his lectures. Having fled an abusive family home to come to Chicago, Pitts decided not to return. He remained in the city, homeless.

Luckily, the University of Chicago had another world-famous logician for Pitts to criticise – Rudolf Carnap. Again, Pitts wrote up notes – this time identifying issues in Carnap's recent book *The Logical Syntax of Language* – and delivered them to Carnap's office at the University of Chicago. Pitts didn't stick around long enough to hear his

---

* At least that's what it looked like at the time … More on this later.

reaction, but Carnap, impressed, eventually chased down Pitts, whom he referred to as 'that newsboy who understood logic'. On this occasion, the philosopher he critiqued actually did get Pitts to work with him. Though he never officially enrolled, Pitts functioned effectively as a graduate student for Carnap and fraternised with a group of scholars who were interested in the mathematics of biology.

Warren McCulloch's interest in philosophy took a more traditional form. Born in New Jersey, he studied the subject (along with psychology) at Yale and read many of the greats. He was most enamoured with Immanuel Kant and Gottfried Leibniz (whose ideas were very influential for Russell), and he read the *Principia* at the age of 25. But, despite the beard upon his long face, McCulloch was not a philosopher – he was a physiologist. He attended medical school in Manhattan and then went on to observe the panoply of ways in which the brain can break as a neurology intern at Bellevue Hospital and at the Rockland State Hospital psychiatric facility. In 1941, he joined the University of Illinois at Chicago as the director of the laboratory for basic research in the department of psychiatry.

As with all great origin stories, there are conflicting accounts of how McCulloch and Pitts met. One claims that it happened when McCulloch spoke in front of a research group Pitts was a part of. Another story is that Carnap introduced them. Finally, a contemporary of the two men, Jerome Lettvin, claimed he introduced them and that all three bonded over a mutual love of Leibniz. In any case, by 1942, the 43-year-old McCulloch and his wife had taken the 18-year-old Pitts into their home, and the two men were spending evenings drinking whisky and discussing logic.

The wall between 'mind' and 'body' was strong among scientists in the early twentieth century. The mind was considered internal and intangible; the body, including the brain, was physical. Researchers on either side of this wall toiled diligently, but separately, at their own problems. Biologists, as we saw in the last chapter, were working hard to uncover the physical machinery of neurons: using pipettes and electrodes and chemicals to sort out what causes a spike and how. Psychiatrists, on the other hand, were attempting to uncover the machinery of the mind through lengthy sessions of Freudian psychoanalysis. Few on either side would attempt a glance over the wall at the other. They spoke separate languages and worked towards different goals. For most practitioners, the question of how neural building blocks could create the structure of the mind was not just unanswered, it was unasked.

But McCulloch, from as early on as his time at medical school, had immersed himself in a crowd of scientists who did care about this question and allowed him the space to think about it. Eventually, through his physiological observations, he came up with a hunch. He saw in the emerging concepts of neuroscience a possible mapping to the notions of logic and computation he so adored in philosophy. To think of the brain as a computing device following the rules of logic – rather than just a bag of proteins and chemicals – would open the door to understanding thought in terms of neural activity.

Analytical skill, however, was not where McCulloch excelled. Some who knew him say he was too much of a romantic to be held down by such details. So, despite years of toying with these ideas in his mind and in conversation (even as a Bellevue intern he was accused of 'trying to write an equation for the working of the brain'), McCulloch struggled with several technical issues of how to enact them. Pitts, however, was comparably unfazed by the analytical. As soon as he spoke with him about it, Pitts saw what approaches were needed to formally realise McCulloch's intuitions. Not long after they met, one of the most influential papers on computation was written.

'A logical calculus of the ideas immanent in nervous activity' was published in 1943. The paper is 17 pages long with many equations, only three references (one of which is to the *Principia*) and a single figure consisting of little neural circuits drawn by McCulloch's daughter.[*]

The paper begins by reviewing the biology of neurons that was known at the time: neurons have cell bodies and axons; two neurons connect when the axon of the first meets the body of the second; through this connection one neuron provides input to the other; a certain amount of input is needed for a neuron to fire; a cell either fires a spike or it doesn't – no half spikes or in-between spikes; and the input from some neurons – inhibitory neurons – has the power to prevent a cell from spiking.

---

[*] The use of the word 'circuit' here differs from that in the last chapter. In addition to its meaning as an electrical circuit, neuroscientists also use the word to refer to a group of neurons connected in a specific way.

McCulloch and Pitts go on to explain how these biological details are congruent with Boolean logic. The core of their claim is that the activity state of each neuron – either firing or not – is like the truth value of a proposition – either true or false. In their own words, they 'conceive of the response of any neuron as factually equivalent to a proposition which proposed its adequate stimulus'.

By 'its adequate stimulus' they are referring to something about the world. Imagine a neuron in the visual cortex whose activity represents the statement 'the current visual stimulus looks like a duck'. If that neuron is firing, that statement is true; if the neuron is not firing, it is false. Now imagine another neuron, in the auditory cortex, that represents the statement 'the current auditory stimulus is quacking like a duck'. Again, if this neuron is firing, that statement is true, otherwise it is false.

Now we can use the connections between neurons to enact Boolean operations. For example, by giving a third neuron inputs from both of these neurons, we could implement the rule 'if it looks like a duck *and* it quacks like a duck, it's a duck'. All we have to do is build the third neuron such that it will only fire if both of its input neurons are firing. That way, both 'looks like a duck' and 'quacks like a duck' have to be true in order for the conclusion represented by the third neuron ('it's a duck') to be true.

This describes the simple circuit needed to implement the Boolean operation 'and'. McCulloch and Pitts in their paper show how to implement many others. To implement 'or' is very similar, however the strength of the connections from each neuron must be so strong that one input alone is enough to make the output neuron fire. In this case, the 'is a duck' neuron would fire if the 'looks like a duck' neuron *or* the 'quacks like a duck' neuron (or both) were firing. The authors even show how to string together multiple Boolean operations. For example, to implement a statement like 'X and not Y', the neuron representing X connects to an output neuron with a strength enough to make it fire. But the neuron representing Y *inhibits* the output neuron, meaning it prevents it from firing. This way, the output neuron will only fire if the X-representing neuron *is* firing and the Y-representing neuron is *not* (see Figure 4).

These circuits, which are meant to represent what networks of real neurons can do, became known as *artificial* neural networks.

The ability to spot logic at play in the interactions of neurons came from McCulloch's discerning eye. As a physiologist, he knew that neurons were more complex than his simple drawings and equations suggested. They had membranes, ion channels and forking paths of

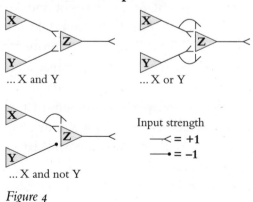

**If neuron Z needs 2 inputs to fire,
it will represent...**

... X and Y

... X or Y

... X and not Y

Input strength

—< = +1

—• = −1

Figure 4

dendrites. But the theory didn't need their full complexity. So, like an impressionist painter using only the necessary strokes, he intentionally highlighted only the elements of neural activity required for the story he wanted to tell. In doing so, he demonstrated the artistry inherent to model-building; it is a subjective and creative process to decide which facts belong in the foreground.

The radical story that McCulloch and Pitts told with their model – that neurons were performing a logical calculus – was the first attempt to use the principles of computation to turn the mind–body problem into a mind–body connection. Networks of neurons were now imbued with all the power of a formal logical system. Like a chain of falling dominoes, once certain truth values entered into a neural population (say, via sensory organs), a cascade of interactions could deduce the truth value of new and different statements. This meant a population of neurons could carry out endless computations: interpreting sensory inputs, developing conclusions, forming plans, reasoning through arguments, performing calculations and so on.

With this step in their research, McCulloch and Pitts advanced the study of human thought and, at the same time, kicked it off its throne. The 'mind' lost its status as mysterious and ethereal once it was brought down to solid ground – that is, once its grand abilities were reduced to the firing of neurons. To adapt a quote from Lettvin, the brain could now be thought of as 'a machine, meaty and miraculous, but still a machine'. More boldly still, McCulloch's student Michael Arbib later remarked that the work 'killed dualism'.

Russell was known to lament that, despite the 20 years put into it and the impact it had on logicians and philosophers, the *Principia* had little effect on practising mathematicians. Its new take on the foundations of mathematics simply didn't seem to mean much to those doing mathematics; it didn't change their day-to-day work. The same could be said of McCulloch and Pitts' discovery for neuroscientists of the time. Biologists, physiologists, anatomists – the scientists doing the labour of physically mining neurons for the details of their workings – didn't take much from the theory. This was in part because it wasn't obvious what experiments should follow from it. But it may also have stemmed from the very technical notation in the paper and its less-than-inviting writing style. In a review on nerve conduction written three years later, the author refers to the McCulloch-Pitts paper as 'not for the lay reader' and remarks that if this style of work is to be useful, it's necessary for 'physiologists to familiarise themselves with mathematical technology, or for mathematicians to elaborate at least their conclusions in a less formidable language'. The wall between mind and body may have come down, but the one between biologist and mathematician stood strong.

There was a separate group of people – a group with the requisite technical know-how – who did take an interest in the logical calculus of neurons. In the post-war era, a series of meetings hosted by the philanthropic Macy Foundation brought together biologists and technologists, many of whom wished to use biological findings to build brain-like machines. McCulloch was an organiser of these meetings, and fellow attendees included the 'father of cybernetics' Norbert Wiener and John von Neumann, the inventor of the modern computer architecture, who was directly inspired in its design by the McCulloch–Pitts neurons. As Lettvin described it 40 years later: 'The whole field of neurology and neurobiology ignored the structure, the message and the form of McCulloch and Pitts' theory. Instead, those who were inspired by it were those who were destined to become the aficionados of a new venture, now called Artificial Intelligence.'

★ ★ ★

The *Navy last week demonstrated the embryo of an electronic computer named the Perceptron which, when completed in about a year, is expected to be the first non-living mechanism able to 'perceive, recognise, and identify its surroundings without human training or control.' [...]*

"'Dr. Frank Rosenblatt, research psychologist at the Cornell Aeronautical Laboratory, Inc., Buffalo, NY, designer of the Perceptron, conducted the demonstration. The machine, he said, would be the first electronic device to think as the human brain. Like humans, Perceptron will make mistakes at first, 'but it will grow wiser as it gains experience', he said.

This summary, from an article entitled 'Electronic "brain" teaches itself', appeared in the 13 July 1958 edition of the *New York Times*, opposite a letter to the editor about the ongoing debate on whether smoking causes cancer. Frank Rosenblatt, the 30-year-old architect of the project, was reaching beyond his training in experimental psychology to build a computer meant to rival the most advanced technology at the time.

The computer in question was taller than the engineers who operated it and about twice as long. It was covered on either end in various control panels and readout mechanisms. Rosenblatt requested the services of three 'professional people' and an associated technical staff for 18 months to build it, and the estimated cost was $100,000 (around $870,000 today). The word 'perceptron', defined by Rosenblatt, is a generic term for a certain class of devices that can 'recognise similarities or identities between patterns of optical, electrical or tonal information'. The Perceptron – the computer that was built in 1958 – was thus technically a subclass known as a 'photoperceptron' because it took as its input the output of a camera mounted on a tripod at one end of the machine.

The Perceptron was, just like the models introduced in the McCulloch–Pitts paper, an artificial neural network. It was a simplified replica of what real neurons do and how they connect to each other. But rather than remaining a mathematical construct that exists only as the ink of equations on a page, the Perceptron was physically realised. The camera provided 400 inputs to this network in the form of a 20x20 grid of light sensors. Wires then randomly connected the output of these sensors to 1,000 'association units' – small electrical circuits that summed up their inputs and switched to 'on' or 'off' as a result, just like a neuron. The output of these association units became the input to the 'response units', which themselves could be 'on' or 'off'. The number of response units was equal to the number of mutually exclusive categories to which an image could belong. So, if the Navy wanted to use the Perceptron, say, to determine if a jet was present in an image or not, there would be two response units: one for jet and one for no jet. At the end of the machine opposite the camera was a set of light bulbs that let the engineer know which of the response units was active – that is, which category the input belonged to.

Implementing an artificial neural network this way was large and cumbersome, full of switches, plugboards and gas tubes. The same network made up of real neurons would be smaller than a grain of sea salt. But achieving this physical implementation was important. It meant that theories of how neurons compute could actually be tested in the real world on real data. Whereas the McCulloch-Pitts work was about proving a point in theory, the Perceptron put it into practice.

Another important difference between the Perceptron and the McCulloch–Pitts network was that, as Rosenblatt told the *New York Times*, the Perceptron learns. In the McCulloch and Pitts paper, the authors make no reference to how the connectivity between the neurons comes to be. It is simply defined according to what logical function the network needs to carry out and it stays that way. For the Perceptron to learn, however, it must modify its connections.[*] In fact, the Perceptron derives all its functionality from changing its connection strengths until they are just right.

The type of learning the Perceptron engages in is known as 'supervised' learning. By providing pairs of inputs and outputs – say, a series of pictures and whether

---

[*] More on how learning – and memory – relies on a change in connections in the next chapter.

they each contain a jet or not – the Perceptron learns to make this decision on its own. It does so by changing the strength of the connections – also known as the 'weights' – between the association units and the readouts.

Specifically, when an image is provided to the network, it activates units first in the input layer, then in the association layer, and finally in the readout layer, indicating the network's decision. If the network gets the classification wrong, the weights change according to these rules:

1. If a readout unit is 'off' when it should be 'on', the connections from the 'on' association units to that readout unit are *strengthened*.
2. If a readout unit is 'on' when it should be 'off', the connections from the 'on' association units to that readout unit are *weakened*.

By following these rules, the network will start to correctly associate images with the category they belong to. If the network can learn to do this well, it will stop making errors and the weights will stop changing.

This procedure for learning was, in many ways, the most remarkable part of the Perceptron. It was the conceptual key that could open all doors. Rather than needing to tell a computer exactly how to solve a problem, you need only show it some examples of that problem solved. This had the potential to revolutionise computing and Rosenblatt was not shy in saying so. He told the *New York Times* that Perceptrons would 'be able to recognise people and call out their names' and 'to hear speech in one language and instantly translate it to speech or writing in another language'. He also added that 'it would be possible to build Perceptrons that could reproduce themselves on an assembly line and which would be "conscious" of their existence'. This was a bold statement, to say the least, and not everyone was happy with Rosenblatt's public bravado. But the spirit of the claim – that a computer that could learn would expedite the solving of almost any problem – rang true.

The power of learning, however, came with a price. Letting the system decide its own connectivity effectively divorced these connections from the concept of Boolean operators. The network *could* learn the connectivity that McCulloch and Pitts had identified as required for 'and', 'or', *etc.* But there was no requirement that it does, nor any need to understand the system in this light. Furthermore, while the association units in the Perceptron machine were designed to be only 'on' or 'off', the learning rule doesn't actually require that they be this way. In fact, the activity level of these artificial neurons could be any positive number and the rule would still work.* This makes the system more flexible, but without a binary 'on'-'off' response it makes it harder to map the activity of these units to the binary truth values of propositions. Compared with the crisp and clear logic of the McCulloch-Pitts networks, the Perceptron was an uninterpretable mess. But it worked. Interpretability was sacrificed for ability.

---

* This can be thought of as representing the *rate* of spiking of a neuron, rather than if the neuron is emitting a spike or not. Using this type of artificial neuron only requires a small modification to the learning procedure.

The Perceptron machine and its associated learning procedure became a popular object of study in the burgeoning field of artificial intelligence. When it made the transition from a specific physical object (the Perceptron) to an abstract mathematical concept (the perceptron algorithm) the separate input and association layers were done away with. Instead, input units representing incoming data connected directly to the readout units and, through learning, these connections changed to make the network better at its task. How and what the perceptron in this simplified form could learn was studied from every angle. Researchers explored its workings mathematically using pen and paper, or physically by building their own perceptron machines, or – when digital computers finally became available – electronically by simulating it.

The perceptron generated hope that humans could build machines that learn like we do; in this way it put the prospect of artificial intelligence within reach. Simultaneously, it provided a new way of understanding our own intelligence. It showed that artificial neural networks could compute without abiding by the strict rules of logic. If the perceptron could perceive without the use of propositions or operators, it follows that each neuron and connection in the brain needn't have a clear role in terms of Boolean logic either. Instead, the brain could be working in a sloppier way, wherein, like the perceptron, the function of a network is distributed across its neurons and emerges out of the connections between them. This new approach to the study of the brain became known as 'connectionism'.

The work of McCulloch and Pitts was an important stepping stone. As the first demonstration of how networks of neurons could think, it was responsible for getting neuroscience away from the shores of pure biology and into the sea of computation. This fact, rather than the veracity of its claims, is what earns it its place in history. The intellectual ancestor of McCulloch and Pitts' work, the *Principia Mathematica*, could be said to have suffered a similar fate. In 1931, German mathematician Kurt Gödel published 'On formally undecidable propositions of *Principia Mathematica* and related systems'. This paper took the *Principia Mathematica* as a starting point to show why its very goal – to explain all of mathematics from simple premises – was impossible to achieve. Russell and Whitehead had not, in fact, done what they believed they did.* Gödel's findings became known as the 'incompleteness theorem' and had a revolutionary effect on mathematics and philosophy. An effect that stemmed, in part, from Russell and Whitehead's failed attempt.

Russell and McCulloch were able to take the failings of their respective works in their stride. Pitts, on the other hand, was made of finer cloth. The realisation that the brain was not enacting the beautiful rules of logic tore him apart.† This, along with pre-existing mental struggles and the end of a relationship with an important mentor, drove him to drink and experiment with other

---

* The cracks in the *Principia*'s foundation were noticeable even when it was published. Some of the 'basic' premises it had to assume were not really very basic and were hard to justify.
† This realisation came even more directly from a study on the frog brain that Pitts was involved with. More on that study in Chapter 6.

drugs. He became erratic and delirious; he burned his work and withdrew from his friends. He died from the impacts of liver disease in 1969 – the same year McCulloch died. McCulloch was 70; Pitts was 46.

★ ★ ★

The cerebellum is a forest. Folded up neatly near where the spinal cord enters the skull, this bit of the brain is thick with different types of neurons, like different species of trees, all living in chaotic harmony (see Figure 5). The Purkinje cells are large, easily identified and heavily branched: from the body of these cells, dendrites stretch up and away, like a thousand alien hands raised in prayer. The granule cells are numerous and small – with cell bodies less than half the size of the Purkinje's – but their reach is far. Their axons initially grow upwards, in parallel with the Purkinje cells' dendrites. They then make a sharp right turn to run directly through the branches of the Purkinje cells, like power lines through treetops. This is where the granule cells make contact with the Purkinje cells: each Purkinje cell gets input from hundreds of

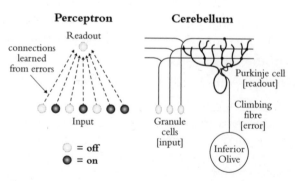

Figure 5

thousands of granule cells. Climbing fibres are axons that follow a longer path on their way to the Purkinje cells. These axons come from cells in a different brain region – the inferior olive – from which they navigate all the way to the bottom of the Purkinje cell bodies and creep up around them. Winding their way around the base of the Purkinje cell dendrites like ivy, the climbing fibres form connections. Unlike the granule cells, only a single climbing fibre targets each Purkinje cell. In the cerebellar landscape, Purkinje cells are thus central. They have scores of granule cells imposing on them from the top and a small yet precise set of climbing fibres closing in on them from the bottom.

In its twisty, organic way, the circuitry of the cerebellum possesses an organisation and precision unbefitting of biology. It was in this biological wiring that James Albus, a PhD student in electrical engineering working at NASA, saw the principles of the perceptron at play.

The cerebellum plays a crucial role in motor control; it helps with balance, coordination, reflexes and more. One of the most widely studied of its abilities is eye-blink conditioning. This is a trained reflex that can be found in everyday life. For example, if a determined parent or roommate tries to get you out of bed in the morning by pulling open the curtains, you'll instinctively close your eyes in response to the sunlight. After a few days of this, simply the sound of the curtains being opened may be enough to make you blink in anticipation.

In the lab, this process is studied in rabbits and the intruding sunlight is replaced by a small puff of air on the eye (just annoying enough to ensure they'll want to avoid it). After several trials of playing a sound (such as a short blip of a pure tone) and following it by this little air puff, the rabbit eventually learns to close its eyes immediately upon hearing the tone. Play the animal a new sound (for example, a clapping noise) that hasn't been paired with air puffs and it won't blink. This makes eye-blink conditioning a simple classification task: the rabbit has to decide if a sound it hears is indicative of an upcoming air puff (in which case the eyes should close) or if it is a neutral noise (in which case they can remain open). Disrupt the cerebellum and rabbits can't learn this task.

The Purkinje cells have the power to close the eyes. Specifically, a dip in their normally high firing rate will, via connections the Purkinje cells send out of this area, cause the eyes to shut. Based on this anatomy, Albus saw their place as the readout – that is, they indicate the outcome of the classification.

The perceptron learns via supervision: it needs inputs and labels for those inputs to know when it has erred. Albus saw these two functions in the two different types of connections on to Purkinje cells. The granule cells pass along sensory signals; specifically, different granule cells fire depending on which sound is being played. The climbing fibres tell the cerebellum about the air puff; they fire when this annoyance is felt. Importantly, this means the climbing fibres signal an error. They indicate that the animal made a mistake in not closing its eyes when it should have.

To prevent this error, the connections from the granule cells to the Purkinje cells need to change. In particular, Albus anticipated that any granule cells that were active before the climbing fibre was active (*i.e.*, before an error), should weaken their connection to the Purkinje cell. That way, the next time those granule cells fire – *i.e.*, the next time the same sound is played – they *won't* cause firing in the Purkinje cells. And that dip in Purkinje cell firing *will* cause the eyes to close. Through this changing of connection strengths, the animal learns from its past mistakes and avoids future air puffs to the eye.

In this way, the Purkinje cell acts like a president advised by a cabinet of granule cell advisors. At first the Purkinje cell listens to all of them. But if it's clear that some are providing bad advice – that is, their input is followed by negative news delivered by the climbing fibre – their influence over the Purkinje cell will fade. And the Purkinje cell will act better in the future. It is a process that directly mirrors the perceptron learning rule.

When Albus proposed this mapping between the perceptron and the cerebellum in 1971,* his prediction about how the connections between granule cells and Purkinje cells should change was just that – a prediction. No one had directly observed this kind of learning in the cerebellum. But by the mid-1980s, evidence had piled up in Albus' favour. It became clear that the strength of the connection between a granule cell and a Purkinje cell does decrease after an error. The particular molecular mechanisms of this process have even been revealed. We now know that granule cell inputs cause a receptor in

---

* The mapping is sometimes referred to as the 'Marr-Albus-Ito' theory of motor learning, named also after David Marr and Masao Ito, who both put forth similar models of how the cerebellum learns.

the membrane of the Purkinje cell to respond, effectively tagging which granule cell inputs were active at a given time. If a climbing fibre input comes later (during an air puff), it causes calcium to flood into the Purkinje cell. The presence of this calcium signals to all the tagged connections to decrease their strength. Patients with fragile X syndrome – a genetic disorder that leads to intellectual disabilities – appear to be missing a protein that regulates this connection from the granule cells on to the Purkinje cell. As a result, they have trouble learning tasks like eye-blink conditioning.

The perceptron, with its explicit rules of how learning should proceed in a neural network, offered clear testable ideas for neuroscientists to hunt for – and find – in the brain. In doing so, it was able to connect science across scales. The smallest physical detail – calcium ions moving through the inside of a neuron, for example – inherits a much larger meaning in light of its role in computation.

★ ★ ★

The reign of the perceptron was cut short in 1969. And with a twist of Shakespearean irony, it was its namesake that killed it.

Perceptrons was written by Marvin Minsky and Seymour Papert, both mathematicians at the Massachusetts Institute of Technology. The book was subtitled *An Introduction to Computational Geometry* and had a simple abstract design on the cover. Minsky and Papert were drawn to write about the topic of perceptrons out of appreciation for Rosenblatt's invention and a desire to explore it further.

In fact, Minsky and Papert met at a conference where they were presenting similar results from their explorations into how the perceptron learns.

Papert was a native of South Africa with full cheeks, a healthy beard and not one, but *two*, PhDs in mathematics. He had a lifelong interest in education and how it could be transformed by computing. Minsky was less than a year older than Papert, with sharper features and large glasses. A New York native, he attended the Bronx High School of Science with Frank Rosenblatt; he was also mentored by McCulloch and Pitts.

Minsky and Papert shared with McCulloch and Pitts the compulsive desire to formalise thinking. True advances in understanding computation, they believed, came from mathematical derivations. All the empirical success of the perceptron – whatever computing it was able to carry out or categories it was able to learn – meant next to nothing without a mathematical understanding of why and how it worked.

At this time, the perceptron was attracting a lot of attention – and money – from the artificial intelligence community. But it wasn't attracting the kind of mathematical scrutiny Minsky and Papert yearned for. The two were thus explicitly motivated to write their book by a desire to increase the rigour around the study of perceptrons – but also, as Papert later acknowledged, by some desire to decrease the reverence for them.*

The pages of *Perceptrons* are comprised mainly of proofs, theorems and derivations. Each contributes to a

---

* The particular words Papert used to describe his feelings about Perceptron-mania at the time were 'hostility' and 'annoyance'.

story about the perceptron: defining what it is, what it can do and how it learns. Yet from the publication of these some 200 pages – a thorough exploration of the ins and outs of the perceptron's workings – the message the community received was largely about its limitations. This is because Minsky and Papert had shown, conclusively, that certain simple computations were impossible for the perceptron to do.

Consider a perceptron that has just two inputs, and each input can be 'on' or 'off'. We want the perceptron to report if the two inputs are the same: to respond yes (*i.e.*, have its readout unit be on) if both inputs are on *or* if both inputs are off. But if one is on and the other is off, the readout unit should be off. Like sorting socks out of the laundry, the perceptron should only respond when it sees a matching pair.

To make sure the readout doesn't fire when only one input is on, the weights from each input need to be sufficiently low. They could, for example, each be half the amount needed to make the readout turn on. This way, when both are on, the readout *will* fire and it won't fire when only one input is on. In this setup the readout is responding correctly for three of the four possible input conditions. But in the condition where both inputs are off, the readout will be off – an incorrect classification.

As it turns out, no matter how much we fiddle with connection strengths, there is no way to satisfy all the needs of the classification at once. The perceptron simply cannot do it. And the problem with that is that no good model of the brain – or promising artificial intelligence – should fail at a task as simple as deciding if two things are the same or not.

Albus, whose paper on the cerebellum was published in 1971, knew of the limitations of the perceptron and knew that, despite these limitations, it was still powerful enough to be a model of the eye-blink conditioning task. But a model of the whole human brain, as Rosenblatt promised? Not possible.

The portrait that Minsky and Papert painted forced researchers to see the perceptron's powers clearly. Prior to the book, researchers were able to explore what the perceptron could do blindly, with the hope that the limits of its abilities were still far off, if they existed at all. Once the contours were put in stark relief, however, there was no denying that these boundaries existed, and that they existed much closer than expected. In truth, all this amounted to was an understanding of the perceptron – exactly what Minsky and Papert set out to do. But the end of ignorance around the perceptron meant the end of excitement around it as well. As Papert put it: 'Being understood can be a fate as bad as death.'

The period that followed the publication of *Perceptrons* is known as the 'dark ages' of connectionism. It was marked by significant decreases in funding to the research programmes that had grown out of Rosenblatt's initial work. The neural network approach to building artificial intelligence was snuffed out. All the excessive promises, hopes and hype had to be retracted. Rosenblatt himself died tragically in a boating accident two years after the book was published and the field he helped build remained dormant for more than 10 years.

But if the hype around the perceptron was excessive and ill informed, so too was the backlash against it. The limitations in Minsky and Papert's book were true: the perceptron in the form they were studying it was incapable of many things. But it didn't need to keep that form. The same-or-not problem, for example, could be easily solved by adding an additional layer of neurons between the input and the readout. This layer could be composed of two neurons, one with weights that make it fire when both inputs are on and the other with weights that make it fire when both inputs are off. Now the readout neuron, which gets its input from these middle neurons, just needs to be active if one of the middle neurons is active.

'Multi-layer perceptrons', as these new neural architectures were called, had the potential to bring connectionism back from the dead.* But before a full resurrection was possible, one problem had to be solved: learning. The original perceptron algorithm provided the recipe for setting the connections between the input neurons and the readout neurons – that is, the learning rule was designed for a two-layered network. If the new breed of neural networks was going to have three, four, five or more layers, how should the connections between all those layers be set? (see Figure 6) Despite all the good features of the perceptron learning rule – its simplicity, the proof that it could work, the fact that it had been spotted in the wild of the cerebellum – it was unable to answer this question. Knowing that a multi-layer perceptron *could* solve more complex problems was not enough to deliver

---

* Technically they weren't 'new'. Minsky and Papert do reference multi-layer perceptrons in their book. However, they were dismissive about the potential powers of these devices and, unfortunately for science, did not encourage their further study.

on the grand promises of connectionism. What was needed was for it to *learn* to solve those problems.

★ ★ ★

The Easter Sunday of the connectionist revival story came in 1986. The paper 'Learning representations by back-propagating errors', written by two cognitive scientists from the University of California San Diego, David Rumelhart and Ronald Williams, and a computer scientist from Carnegie Mellon, Geoffrey Hinton, was published on 9 October in the journal *Nature*. It presented a solution to the exact problem the field had: how to train multi-layer artificial neural networks. The learning algorithm in the paper, called 'backpropagation', became widely used by the community at the time. And it remains to this day the dominant way in which artificial neural networks are trained to do interesting tasks.

The original perceptron's learning rule works because, with only two layers, it's easy to see how to fix what's gone wrong. If a readout neuron is off when it should be on, connections going from the input layer to that neuron

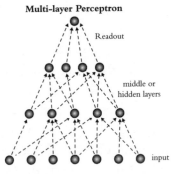

Figure 6

need to get stronger and vice versa. The relationship between these connections and the readout is thus clear. The backpropagation algorithm has a more difficult problem to solve. In a network with many layers between the input and readout, the relationships between all these connections and the readout aren't as clear. Now instead of a president and his or her advisors, we have a president, their advisors, and the employees of those advisors. The amount of trust an advisor has in any given employee – *i.e.*, the strength of the connection from that employee to the advisor – will certainly have an impact, ultimately, on what the president does. But this impact is harder to directly see and harder to fix if the president feels something is going wrong.

What was needed was an explicit way to calculate how any connection in the network would impact the readout layer. As it turns out, mathematics offers a neat way to do this. Consider an artificial neural network with three layers: input, middle and readout. How do the connections from the input to the middle layer impact the readout? We know the activity of the middle layer is a result of the activities of the input neurons and the weights of their connections to the middle layer. With this knowledge, writing an equation for how these weights affect the activity at the middle layer is straightforward. We also know that the readout neurons follow the same rule: their activity is determined by the activities of the middle neurons and the weights connecting the middle neurons to the readout. Therefore, an equation describing how these weights impact the readout is also easy to get. The only thing left to do is find a way to string these two equations together. That way we'll have an equation that tells us directly how the connections from the input to the middle layer impact the readout.

When forming a train in the game of dominoes, the number on the end of one tile needs to match the number on the start of another in order for them to connect. The same is true for stitching together these equations. Here, the common term that connects the two equations is the activity of the middle layer: this activity both determines the activity of the readout and is determined by the input-to-middle connections. After joining these equations via the middle layer activity, the impact of the input-to-middle layer connections on the readout can be calculated directly. And this makes it easy to figure out how those connections should change when the readout is wrong. In calculus, this linking together of relationships is known as the 'chain rule' and it is the core of the backpropagation algorithm.

The chain rule was discovered over 200 years ago by none other than the idol of McCulloch and Pitts, philosopher and polymath Gottfried Leibniz. Given how useful the rule is, its application to the training of multi-layer neural networks was no surprise. In fact, the backpropagation algorithm appears to have been invented at least three separate times before 1986. But the 1986 paper was part of a perfect storm that ensured its findings would spread far and wide. The first reason for this was the content of the paper itself. Not only did it show that neural networks could be trained this way, it also analysed the workings of networks trained on several cognitive tasks, such as understanding relations on a family tree. Another component of the success was the increase in computational power that came in the 1980s; this was important for making the training of multi-layer neural networks practically possible for researchers. Finally, the same year the paper was published, one of its authors, Rumelhart, also published a book on connectionism that included the backpropagation algorithm. That book – written with a different Carnegie Mellon professor, James McClelland – went on to sell an estimated 40,000 copies by the mid-1990s. Its title, *Parallel Distributed Processing*, lent its name to the entire research agenda of building artificial neural networks in the late 1980s and early 1990s.

For somewhat similar reasons, the story of artificial neural networks took an even more dramatic turn roughly a decade into the new millennium. The heaps of data accumulated in the internet age united with the computational power of the twenty-first century to supercharge progress in this field. Networks with more and more layers could suddenly be trained on more and more complex tasks. Such scaled-up models – referred to as 'deep neural networks' – are currently transforming artificial intelligence and neuroscience alike.

The deep neural networks of today are based on the same basic understanding of neurons as those of McCulloch and Pitts. Beyond that base inspiration, though, they don't aim to directly replicate the human brain. They aren't trying to mimic its structure or anatomy, for example.* But they do aim to mimic human behaviour and they're getting quite good at it. When

---

* With the exception of deep neural networks that are built to understand images, which we will hear all about in Chapter 6.

Google's popular language translation service started using a deep neural network approach in 2016, it reduced translation errors by 50 per cent. YouTube also uses deep neural networks to help its recommendation algorithm better understand what videos people want to see. And when Apple's voice assistant Siri responds to a command, it is a deep neural network that is doing the listening and the speaking.

In total, deep neural networks can now be trained to find objects in images, play games, understand preferences, translate between different languages, turn speech into written words and turn written words into speech. Not unlike the original Perceptron machine, the computers these networks run on fill up rooms. They're located in server centres across the globe, where they hum away processing the world's image, text and audio data. Rosenblatt may have been pleased to see that some of his grand promises to the *New York Times* were indeed fulfilled. They just required a scale nearly a thousand times what he had available at the time.

The backpropagation algorithm was necessary to boost artificial neural networks to the point where they could reach near-human levels of performance on some tasks. As a learning rule for neural networks, it really works. Unfortunately, that doesn't mean it works like the brain. While the perceptron learning rule was something that could be seen at play between real neurons, the backpropagation algorithm is not. It was designed as a mathematical tool to make artificial neural networks work, not a model of how the brain learns (and its inventors were very clear on that from the start). The reason for this is that real neurons can typically only know about the activity of the neurons they're connected to – not about the activity of the neurons those neurons connect to and so on and so on. For this reason, there is no obvious way for real neurons to implement the chain rule. They must be doing something different.

For some researchers – particularly researchers in the field of artificial intelligence – the artificial nature of backpropagation is no problem. Their goal is to build computers that can think, by whatever means necessary. But for other scientists – neuroscientists in particular – finding the learning algorithm of the brain is paramount. We know the brain is good at getting better; we see it when we learn a musical instrument, how to drive or how to read a new language. The question is how.

Because backpropagation is what we know works, some neuroscientists are starting there. They're checking for signs that the brain is doing something *like* backpropagation, even if it can't do it exactly. Inspiration comes from the success story of finding a perceptron at work in the cerebellum. There, clues were present in the anatomy; the different placement of the climbing fibres and granule cells pointed to a different role for each. Other brain areas display patterns of connectivity which may hint at how they are learning. For example, in the neocortex, some neurons have dendrites that stretch out way above them. Faraway regions of the brain send inputs to these dendrites. Do they carry with them information about how these neurons have impacted those that come after them in the brain's neural network? Could this information be used to change the strength of the network's connections? Both neuroscientists and artificial intelligence researchers hold out hope that the brain's version of backpropagation will be found and that, when it is, it can be copied to create algorithms that learn even better and faster than today's artificial neural networks.

In their hunt to understand how the mind learns from supervision, modern researchers are doing just what McCulloch did. They're looking at the piles of facts we have about the biology of the brain and trying to see in it a computational structure. Today, they are guided in their search by the workings of artificial systems. Tomorrow, the findings from biology will again guide the building of artificial intelligence. This back-and-forth defines the symbiotic relationship between these two fields. Researchers looking to build artificial neural networks can take inspiration from the patterns found in biological ones, while neuroscientists can look to the study of artificial intelligence to identify the computational role of biological details. In this way, artificial neural networks keep the study of the mind and the brain connected.

CHAPTER FOUR

Making and Maintaining Memories

The Hopfield network and attractors

A block of iron at 770°C (1,418°F) is a sturdy grey mesh. Each of its trillions of atoms serves as a single brick in the endless parallel walls and ceilings of its crystalline structure. It is a paragon of orderliness. In opposition to their organised structural arrangement, however, the magnetic arrangement of these atoms is a mess.

Each iron atom forms a dipole – a miniature magnet with one positive and one negative end. Heat unsteadies these atoms, flipping the direction of their poles around at random. On the micro-level this means many tiny magnets each exerting a force in its own direction. But as these forces work against each other, their net effect becomes negligible. When you zoom out, this mass of mini-magnets has no magnetism at all.

As the temperature dips below 770°C, however, something changes. The direction of an individual atom is less likely to switch. With its dipole set in place, the atom starts to exert a constant pressure on its neighbours. This indicates to them which direction they too should be facing. Atoms with different directions vie for influence over the local group until eventually everyone falls in line, one way or the other. With all the small dipoles aligned, the net force is strong. The previously inert block of iron becomes a powerful magnet.

Philip Warren Anderson, an American physicist who won a Nobel Prize working on such phenomena, wrote in a now-famous essay entitled 'More is different' that 'the behaviour of large and complex aggregates of elementary particles, it turns out, is not to be understood in terms of a simple extrapolation of the properties of a few particles'. That is, the collective action of many small particles – organised only through their local interactions – can produce a function not directly possible in any of them alone. Physicists have formalised these interactions as equations and successfully used them to explain the behaviour of metals, gases and ice.

In the late 1970s, a colleague of Anderson's, John J. Hopfield, saw in these mathematical models of magnetism a structure akin to that of the brain. Hopfield used this insight to bring under mathematical control a long-lasting mystery: the question of how neurons make and maintain memories.

★ ★ ★

Richard Semon was wrong.

A German biologist working at the turn of the twentieth century, Semon wrote two lengthy books on the science of memory. They were filled with detailed descriptions of experimental results, theories and a vocabulary for describing memory's impact on 'organic tissue'. Semon's work was insightful, honest and clear – but it contained a major flaw. Just as French naturalist Jean-Baptiste Lamarck believed (in contrast to our current understanding of evolution) that traits acquired by an animal in its lifetime could be passed to its offspring,

Semon proposed that *memories* acquired by an animal could be passed down. That is, he believed that an organism's learned responses to its own environment would arise without instruction in its offspring. As a result of this mistaken intuition, much of Semon's otherwise valuable work was slowly cast aside and forgotten.

Being wrong about memory isn't unusual. Philosopher René Descartes, for example, thought memories were activated by a small gland directing the flow of 'animal spirits'. What's unique about Semon is that, despite the flaw in his work that sentenced him to historical obscurity, one of his contributions remained influential long enough to spawn an entire body of research. This small artefact of his efforts is the 'engram' – a word coined by Semon in *The Mneme* in 1904, and subsequently learned by millions of students of psychology and neuroscience.

At the time Semon was writing, memory had only recently come under scientific scrutiny – and most of the results were purely about memorisation skills, not about biology. For example, people would be trained to memorise pairs of nonsense words (such as 'wsp' and 'niq') and then were tested on their ability to retrieve the second word when prompted with the first. This type of memory, known as *associative* memory, would become a target of research for decades to come. But Semon was interested in more than just behaviour; he wanted to know what changes in an animal's physiology could support such associative memories.

Led by scant experimental data, he broke up the process of creating and recovering memories into multiple components. Finding common words too

vague and overloaded, he created novel terms for these
divisions of labour. The word that would become so
influential, the engram, was defined as 'the enduring
though primarily latent modification in the irritable
substance produced by a stimulus'. Or, to put it more
plainly: the physical changes in the brain that happen
when a memory is formed. Another term, 'ecphory', was
assigned to the 'influences which awake the mnemic
trace or engram out of its latent state into one of
manifested activity'. This distinction between engram
and ecphory (or between the processes that lay a memory
and those that retrieve it) was one of the many conceptual
advances that Semon's work provided. Despite the fact
that his name and most of his language have disappeared
from the literature, many of Semon's conceptual insights
were correct and they form the core of how memories
are modelled today.

In 1950, American psychologist Karl Lashley
published 'In search of the engram', a paper that
solidified the legacy of the word. It also set a rather
dismal tone for the field. The paper was so titled because
the search was all Lashley felt he had accomplished in
30 years of experiments. Lashley's experiments involved
training animals to make an association (for example, to
react in a specific way when shown a circle versus an
'X') or learn a task, such as how to run through a
particular maze. He would then surgically remove
specific brain areas or connection pathways and observe
how behaviour was impacted post-operatively. Lashley
couldn't find any area or pattern of lesions that reliably
interfered with memory. He concluded that memories
must thus somehow be distributed equally across the

brain, rather than in any single area. But based on some calculations about how many neurons could be used for a memory and the number of pathways between them, he was uncertain about how this was possible. His landmark article thus reads as something of a white flag, a surrendering of any attempt to draw conclusions about the location of memory in the face of a mass of inconsistent data. The physical nature of memory remained to Lashley as vexing as ever.

At the same time, however, a former student of Lashley's was developing his own theories on learning and memory.

Donald Hebb, a Canadian psychologist whose early work as a school teacher grew his interest in the mind, was intent on making psychology a biological science. In his 1949 book, *The Organization of Behavior*, he describes the task of a psychologist as 'reducing the vagaries of human thought to a mechanical process of cause and effect'. And in that book, he lays down the mechanical process he believed to be behind memory formation.[*] Overcoming the limited, and sometimes misleading, physiological data available at the time, Hebb came to this principle about the physical underpinnings of learning largely through intuition. Yet it would go on to have huge empirical success. The principle, now known as Hebbian learning, is succinctly described by the phrase 'neurons that fire together wire together'.

---

[*] Jerzy Konorski, a Polish neurophysiologist, published a book with very similar ideas the year before Hebb did. In fact, Konorski anticipated several important findings in neuroscience and psychology. However, the global East–West divide at the time isolated his contributions.

Hebbian learning describes what happens at the small junction between two neurons where one can send a signal to the other, a space called the synapse. Suppose there are two neurons, A and B. The axon from neuron A makes a synaptic connection on to the dendrite or cell body of neuron B (making it the 'pre-synaptic' neuron, and neuron B the 'post-synaptic' neuron, see Figure 7). In Hebbian learning, if neuron A repeatedly fires before neuron B, the connection from A to B will strengthen. A stronger connection means that the next time A fires it will be more effective in causing B to fire. In this way, activity determines connectivity and connectivity determines activity.

Hebb's approach, with its focus on the synapse, situates the engram as both local and global: local because a memory's imprint occurs at the small gap where one neuron meets another, but global because these changes may be happening at synapses all across the brain. It also makes memory the natural consequence of experience: with pliable synapses, any activation of the brain has the potential to leave a trace.

Lashley, a dutiful scientist intent on following the facts, accepted that the engram must be distributed based on his own experiments. But he found no satisfaction in

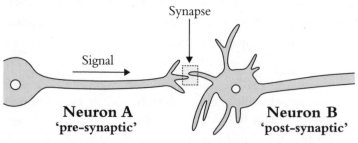

Figure 7

Hebb's solution, which – though an enticing and elegant theory – was based more on speculation than on hard evidence. He turned down Hebb's offer to be a co-author on the work.

Lashley may not have supported Hebb's ideas, but since the publication of his book countless experiments have. Sea slugs – foot-long slimy brown invertebrates with only about 20,000 neurons – became a creature of much study in this area, due to their ability to learn a very basic association. These shell-less slugs have a gill on their backs that, if threatened, can be quickly retracted for safe keeping. In the lab, a short electric shock will cause the gill to withdraw. If such a shock is repeatedly preceded by a harmless light touch, the slug will eventually start withdrawing in response to the touch alone, demonstrating an association between the touch and what's expected to come next. It is the marine critter equivalent of learning to pair 'wsp' with 'niq'. This association was shown, in line with Hebb's theory of learning, to be mediated by a strengthening of the connections between the neurons that represent the touch and those that lead to the gill's response. The change in behaviour was forged through a changing of connections.

Hebbian learning has not just been observed; it's been controlled as well. In 1999, Princeton researchers showed that genetically modifying proteins in the cell membrane that contribute to synaptic changes can control a mouse's capacity for learning. Increasing the function of these receptors enhances the ability of mice to remember objects they've been shown before. Interfering with these proteins impairs it.

It is now established science that experience leads to the activation of neurons and that activating neurons can alter the connections between them. This story is accepted as at least a partial answer to the question of the engram. But, as Semon describes it, the engram itself is only part of the story of memory. Memory also requires remembering. How can this way of depositing memories allow for long-term storage and recall?

★ ★ ★

It was no real surprise that John J. Hopfield became a physicist. Born in 1933 to John Hopfield Sr, a man who made a name for himself in ultraviolet spectroscopy, and Helen Hopfield, who studied atmospheric electromagnetic radiation, Hopfield Jr grew up in a household where physics was as much a philosophy as it was a science. 'Physics was a point of view that the world around us is, with effort, ingenuity and adequate resources, understandable in a predictive and reasonably quantitative fashion,' Hopfield wrote in an autobiography. 'Being a physicist is a dedication to a quest for this kind of understanding.' And a physicist is what he would be.*

Hopfield, a tall and lanky man with an engaging smile, earned his PhD in 1958 from Cornell University. He further emulated his father by receiving a Guggenheim fellowship, using it to study at the Cavendish Laboratory

---

* When Hopfield wrote on an undergraduate admission form that he intended to study 'physics or chemistry', his university advisor – a colleague of his father's – crossed out the latter option saying, 'I don't believe we need to consider chemistry.'

at Cambridge. But even by this stage, Hopfield's enthusiasm for the subject of his PhD – condensed matter physics – was waning. 'In 1968, I had run out of problems … to which my particular talents seemed useful,' he later wrote.

Hopfield's gateway from physics to biology was hemoglobin, a molecule that both serves a crucial biological function as the carrier of oxygen in blood and could be studied with many of the techniques of experimental physics at the time. Hopfield worked on hemoglobin's structure for several years at Bell Labs, but he found his real calling in biology after being invited to a seminar series on neuroscience in Boston in the late 1970s. There he encountered a variegated group of clinicians and neuroscientists, gathered together to address the deep question of how the mind emerges from the brain. Hopfield was captivated.

Mathematically minded as he was, though, Hopfield was dismayed by the qualitative approach to the brain he saw on display. He was concerned that, despite their obvious talents in biology, these researchers, 'would never possibly solve the problem because the solution can be expressed only in an appropriate mathematical language and structure'.* This was a language that physicists had. Hopfield therefore made a point of using his physicist's skillset even as he embarked on a study of memory. In his eyes, certain physicists of the time who made the leap

---

* Hopfield's attitude was not unique. The 1980s found many physicists, bored with their own field, looking at the brain and thinking, 'I could solve that.' After Hopfield's success, this population increased even more.

to biology had immigrated fully, taking on the questions, culture and vocabulary of their new land. He wanted to firmly retain his citizenship as a physicist.

In 1982 Hopfield published 'Neural networks and physical systems with emergent collective computational abilities', which laid out the description and results of what is now known as the Hopfield network. This was Hopfield's first paper on the topic; he was only dipping his toe into the field of neuroscience and yet it made quite the splash.

The Hopfield network (see Figure 8) is a mathematical model of neurons that can implement what Hopfield described as 'content-addressable memory'. This term, coming from computer science, refers to the notion that a full memory can be retrieved from just a small component of it. The network that Hopfield designed for this task is simply composed. It is made only of binary neurons (like the McCulloch-Pitts neurons introduced in the last chapter), which can be either 'on' or 'off'. It is therefore the interactions between these neurons from which the intriguing behaviours of this network emerge.

The Hopfield network is *recurrent*, meaning that each neuron's activity is determined by that of any of the

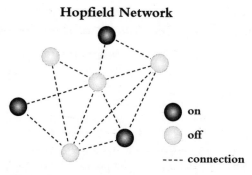

Figure 8

others in the network. Therefore, each neuron's activity serves as both input and output to its neighbours. Specifically, each input a neuron receives from another neuron is multiplied by a particular number – a synaptic weight. These weighted inputs are then added together and compared to a threshold: if the sum is greater than (or equal to) the threshold, the neuron's activity level is 1 ('on'), otherwise it's 0 ('off'). This output then feeds into the input calculations of the other neurons in the network, whose outputs feed back into more input calculations and so on and so on.[*]

Like bodies in a mosh pit, the components of a recurrent system push and pull on each other, with the state of a unit at any given moment determined by those that surround it. The neurons in a Hopfield network are thus just like the atoms of iron constantly influencing each other through their magnetic interactions. The effects of this incessant interaction can be myriad and complex. To predict the patterns these interlocking parts will generate is essentially impossible without the precision of a mathematical model. Hopfield was intimately familiar with these models and their ability to show how local interactions lead to the emergence of global behaviour.

Hopfield found that if the weights between the neurons in his network are just right the network as a

---

[*] While the calculation of an individual neuron's activity in terms of inputs and weights is the same as described for the perceptron in the last chapter, the perceptron is a *feedforward* (not recurrent) network. Recurrence means that the connections can form loops: neuron A connects to neuron B, which connects back to neuron A, for example.

whole can implement associative memory. To understand this, we must first define what counts as a memory in this abstract model. Imagine that each neuron in a Hopfield network represents a single object: neuron A is a rocking chair, neuron B is a bike, neuron C is an elephant, and so on. To represent a particular memory, say that of your childhood bedroom, the neurons that represent all the objects in that room – the bed, your toys, photographs on the wall – should be 'on'; while those that represent objects not in that room – the moon, a city bus, kitchen knives – should be 'off'. The network as a whole is then in the 'your childhood bedroom' activity state. A different activity state – with different sets of neurons 'on' or 'off' – would represent a different memory.

In associative memory, a small input to the network reactivates an entire memory state. For example, seeing a picture of yourself on your childhood bed may activate some of the neurons that represent your bedroom: the bed neurons and pillow neurons, and so on. In the Hopfield network, the connections between these neurons and the ones that represent other parts of the bedroom – the curtains and your toys and your desk – cause these other neurons to become active, recreating the full bedroom experience. Negatively weighted connections between the bedroom neurons and those that represent, say, a local park, ensure that the bedroom memory is not infiltrated by other items. That way you don't end up remembering a swing set next to your closet.

As some neurons turn on and others off, it is their interactivity that brings the full memory into stark relief.

The heavy-lifting of memory is thus done by the synapses. It is the strength of these connections that carries out the formidable yet delicate task of memory retrieval.

In the language of physics, a fully retrieved memory is an example of an *attractor*. An attractor is, in short, a popular pattern of activity. It is one that other patterns of activity will evolve towards, just as water is pulled down a drain. A memory is an attractor because the activation of a few of the neurons that form the memory will drive the network to fill in the rest. Once a network is in an attractor state, it remains there with the neurons fixed in their 'on' or 'off' positions. Always fond of describing things in terms of energy, physicists consider attractors 'low energy' states. They're a comfortable position for a system to be in; that is what makes them attractive and stable.

Imagine a trampoline with a person standing on it. A ball placed anywhere on the trampoline will roll towards the person and stay there. The ball being in the divot created by the person is thus an attractor state for this system. If two people of the same size were standing opposite each other on the trampoline, the system would have two attractors. The ball would roll towards whomever it was initially closest to, but all roads would still lead to an attractor. Memory systems wouldn't be of much use if they could only store one memory, so it is important that the Hopfield network can sustain multiple attractors. The same way the ball is compelled towards the nearest low point on the trampoline, initial neural activity states evolve towards the nearest, most similar memory (see Figure 9). The initial states that lead to a specific memory attractor – for example, the picture

Different inputs reignite different memories

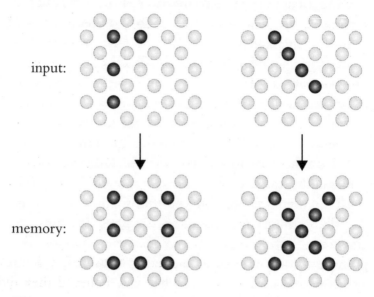

input:

memory:

Figure 9

of your childhood bed that reignites a memory of the whole room or a trip to a beach that ignites the memory of a childhood holiday – are said to be in that memory's 'basin of attraction'.

The Pleasures of Memory is a 1792 poem by Samuel Rogers. Reflecting on the universal journey on which memory can take the mind, he wrote:

> *Lulled in the countless chambers of the brain,*
> *Our thoughts are linked by many a hidden chain.*
> *Awake but one, and lo, what myriads rise!*
> *Each stamps its image as the other flies!*

Rogers' 'hidden chain' can be found in the pattern of weights that reignite a memory in the Hopfield network.

Indeed, the attractor model aligns with much of our intuition about memory. It implicitly addresses the time it takes for memories to be restored, as the network needs time to activate the right neurons. Attractors can also be slightly displaced in the network, creating memories that are mostly correct, with a detail or two changed. And memories that are too similar may simply merge into one. While collapsing memory to a series of zeros and ones may seem an affront to the richness of our experience of it, it is the condensation of this seemingly ineffable process that puts an understanding of it within reach.

In the Hopfield network, how robustly neurons are connected with each other defines which patterns of neural activity form a memory. The engram is therefore in the weights – but how does it get there? How can an experience create just the right weights to make a memory? Hebb tells us that memories should come out of a strengthening of the connections between neurons that have similar activity – and in the Hopfield network that is just how it's done.

The Hopfield network encodes a set of memories through a simple procedure. For every experience in which two neurons are either both active or inactive, the connection between them is strengthened. In this way, the neurons that fire together come to be wired together. On the other hand, for every pattern where one neuron is active and the other is inactive, the connection is weakened.* After this learning

---

* This second part – the idea that connection strength should *decrease* if a pre-synaptic neuron is highly active while the post-synaptic neuron remains quiet – was not part of Hebb's original sketch, but it has since been borne out by experiments.

procedure, neurons that are commonly co-active in memories will have a strong positive connection, those that have opposite activity patterns will have strong negative connections and others will fall somewhere in between. This is just the connectivity needed to form attractors.

Attractors are not trivial phenomena. After all, if all the neurons in a network are constantly sending and receiving inputs, why should we assume their activity would ever settle into a memory state, let alone the *right* memory state? So, to be certain that the right attractors would form in these networks, Hopfield had to make a pretty strange assumption: weights in the Hopfield network are *symmetric*. That means the strength of the connection from neuron A to neuron B is always the same as the strength from B to A. Enforcing this rule offered a mathematical guarantee of attractors. The problem is that the odds of finding a population of neurons like this in the brain are dismal to say the least. It would require that each axon going out of one cell and forming a synapse with another be matched exactly by that same cell sending its axon back, connecting to the first cell with the same strength. Biology simply isn't that clean.

This illuminates the ever-present tension in the mathematical approach to biology. The physicist's perspective, which depends on an almost irrational degree of simplification, is at constant odds with the biology, full as it is of messy, inconvenient details. In this case, the details of the maths demanded symmetric weights in order to make any definitive statement about attractors and thus to make progress on modelling the process of memory. A biologist would likely have dismissed the assumption outright.[*]

Hopfield, with one foot on either side of the mathematics–biology divide, knew to appreciate the perspective of the neuroscientists. To ease their concerns, he showed in his original paper that – even though it couldn't be guaranteed mathematically – networks that allowed asymmetric weights still seemed able to learn and sustain attractors relatively well.

The Hopfield network thus offered a proof of concept that Hebb's ideas about learning could actually work. Beyond that, it offered a chance to study memory mathematically – to quantify it. For example, precisely how many memories can a network hold? This is a question that can only be asked with a precise model of memory in mind. In the simplest version of the Hopfield network, the number of memories depends on the number of neurons in the network. A network with 1,000 neurons, for example, can store about 140 memories; 2,000 neurons can store 280; 10,000 can store 1,400 and so on. If the number of memories remains less than about 14 per cent the number of neurons, each memory will be restored with minimal error. Adding more memories, however, will be like the final addition to a house of cards that causes it to cave in. When pushed past its capacity, the Hopfield network collapses: inputs

---

[*] In fact, when Hopfield presented an early version of this work to a group of neuroscientists, one attendee commented that 'it was a beautiful talk but unfortunately had nothing to do with neurobiology'.

go towards meaningless attractors and no memories are successfully recovered. It's a phenomenon given the appropriately dramatic name 'blackout catastrophe'.*

Precision cannot be evaded; once this estimate of memory capacity is found, it's reasonable to start asking if it aligns with the number of memories we know to be stored by the brain. A landmark study in 1973 showed that people who had been shown more than 10,000 images (each only once and for only a brief period of time) were quite capable of recognising them later on. The 10 million neurons in the perirhinal cortex – a brain region implicated in visual memory – could store this amount of images, but it wouldn't leave much space for anything else. Therefore, there seemed to be a problem with Hebbian learning.

This problem becomes less problematic, however, when we realise that recognition is not recall. That is, a feeling of familiarity when seeing an image can happen without the ability to regenerate that image from scratch. The Hopfield network is remarkable for being capable of the latter, more difficult task – it fully completes a memory from a partial bit of it. But the former task is still important. Thanks to researchers working at the University of Bristol, it's now known that recognition can also be performed by a network that uses Hebbian learning. These networks, when assessed on their ability to label an input as novel or familiar, have a significantly

---

* You may know some people with stories of their own 'blackout catastrophe' after a night of drinking. However, the exact type of memory failure seen in Hopfield networks is not actually believed to occur in humans.

higher capacity: 1,000 neurons can now recognise as many as 23,000 images. Just as Semon so presciently identified, this is an example of an issue that arises from relying on common language to parcel up the functions of the brain. What feels like simply 'memory' to us crumbles when pierced by the scrutiny of science and mathematics into a smattering of different skills.

★ ★ ★

When, in 1953, American doctor William Scoville removed the hippocampus from each side of 27-year-old Henry Molaison's brain, he thought he was helping prevent Molaison's seizures. What Scoville didn't know was the incredible impact this procedure would have on the science of memory. Molaison (more famously known as 'H. M.' in scientific papers to hide his identity until his death in 2008) did find some relief from his seizures after the procedure, but he never formed another conscious memory again. Molaison's subsequent and permanent amnesia initiated a course of research that centred the hippocampus – a curved finger-length structure deep in the brain – as a hub in the memory-formation system. Evading Lashley's troubled search, this is a location that does play a special role in storing memories.

Current theories of hippocampal function go as follows: information about the world first reaches the hippocampus at the dentate gyrus, a region that runs along the bottom edge of the hippocampus. Here, the representation is primed and prepped to be in a form more amenable to memory storage. The dentate gyrus then sends connections on to where attractors are

Hippocampus

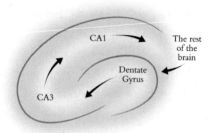

Figure 10

believed to form, an area called CA3; CA3 has extensive recurrent connections that make it a prime substrate for Hopfield network-like effects. This area then sends output to another region called CA1, which acts as a relay station; it sends the remembered information back to the rest of the brain (see Figure 10).

What's interesting about this final step – and what may have muddied Lashley's original findings – is that these projections out to different areas of the brain are believed to facilitate the *copying* of memories. In this way, CA3 acts as a buffer, or warehouse, holding on to memories until they can be transferred to other brain areas. It does so by reactivating the memory in those areas. The hippocampus thus helps the rest of the brain memorise things using the same strategy you'd use to study for a test: repetition. By repeatedly reactivating the same group of neurons elsewhere in the brain, the hippocampus gives those neurons the chance to undergo Hebbian learning themselves. Eventually, their own weights have changed enough for the memory to be safely stored there.* With

---

* This process is believed to occur while you sleep.

his hippocampus gone, Molaison had no warehouse for his experiences, no way to replay his memories back to his brain.

With knowledge of this memory storehouse in the brain, researchers can look into how it works. Particularly, they can look for attractors in it.

In 2005, scientists at University College London recorded the activity of hippocampal cells in rats. The rats got used to being in two different enclosures – a circular one and a square one. Their hippocampal neurons showed one pattern of activity when they were in the circle and a different pattern when they were in the square. The test for attractors came when an animal was placed into a new 'squircle' environment, the shape of which was somewhere in between a circle and a square. The researchers found that if the environment was more square-like the neural activity went to the pattern associated with the square environment; more circle-like and it went to that of the circle. Crucially, there were no intermediate representations in response to intermediate environments, only all circle or all square. This makes the memories of the circle and square environments attractors. An initial input that isn't exactly one or the other is unstable; it gets inescapably driven towards the nearest established memory.

The Hopfield network made manifest the theories of Hebb and showed how attractors – normally studied in physics – could explain the mysteries of memory. Yet Hopfield knew the limitations of bringing mathematics to real brains in real laboratories. He described his own model as a 'mere parody of the complexities of neurobiology'. Indeed, as the creation of a physicist, it lacks all the gooey richness of biology. But as a parody capable of powerful computations, it has offered many insights as well – insights that didn't end with simple storage and recall.

★ ★ ★

You're eating dinner in your kitchen when your roommate comes home. When you see them, you remember that last night you finished a book they had lent you and you want to return it before they leave for a trip the next day. So, you put down your food, head out of the kitchen and go down the hallway. You walk up the stairs, turn, enter your room and think: 'Wait, what am I doing here?'

The sensation is a common one. So much so it's been given a name: 'destinesia' or amnesia about why you've gone to where you are. It's a failure of what's called 'working memory', the ability to hold an idea in mind, even just for the 10 seconds it takes to walk from room to room. Working memory is crucial for just about all aspects of cognition: it's hard to make a decision or work through a plan if you keep forgetting what you're thinking about.

Psychologists have been studying working memory for decades. The term itself was first coined in the 1960 book *Plans and the Structure of Behavior* written by George A. Miller and fellow scientists working at the Center for Advanced Study in the Behavioral Sciences in California. But the concept was explored well before that. Indeed, Miller himself wrote one of the most influential papers on the topic four years previously, in 1956. Perhaps anticipating its fame, Miller gave the paper the cheeky title 'The magical number seven, plus or minus two'. What that magical number refers to is the number of items humans can hold in their working memory at any one time.

An example of how to assess this is to 1) show a participant several coloured squares on a screen; 2) ask them to wait for some time between several seconds to minutes; 3) then show them a second set of coloured squares. The task of the subject is to indicate if the colours of the second set are the same as the colours of the first. People can do well on this task if the number of squares shown remains small, achieving nearly 100 per cent accuracy if only one square is shown. Adding in more squares makes the performance drop and drop, until past seven, it's almost no different to random guessing. Whether seven really is a special value when it comes to this kind of working memory capacity is up for debate; some studies find lower limits, some higher. However, there's no doubt that Miller's paper made an impact and psychologists have worked to characterise nearly every aspect of working memory since, from what can be held in it to how long it can last.

But the question remains of how the brain actually does this: where are working memories stored and in what way? A tried-and-tested method for answering such questions – lesion experiments – pointed to the prefrontal cortex, a large part of the brain just behind the forehead. Whether it was humans with unfortunate injuries or laboratory animals with the area removed, it was clear that damaging the prefrontal cortex reduced working memory substantially. Without it, animals can hardly hold on to an idea for more than a second or two. Thoughts and experiences pass through their minds like water through cupped hands.

With an 'X' marking the spot, neuroscientists then began to dig. Dropping an electrode into the prefrontal cortex of monkeys, in 1971 researchers at the University of California in Los Angeles eavesdropped on the neurons there. The scientists, Joaquin Fuster and Garrett Alexander, did this while the animals performed a task similar to the colour memory test. These tests are known as 'delayed response' tasks because they include a delay period wherein the important information is absent from the screen and must thus be held in memory. The question was: what are neurons in the prefrontal cortex doing during this delay?

Most of the brain areas responsible for vision have a stereotyped response to this kind of task: the neurons respond strongly when the patterns on the screen initially show up and then again when they reappear after the delay, but during the delay period – when no visual inputs are actually entering the brain – these areas are mostly quiet. Out of sight really does mean out of mind for these neurons. What Fuster and Alexander found, however, was that cells in the prefrontal cortex were different. The neurons there that responded to the visual patterns kept firing even after the patterns disappeared; that is, they maintained their activity during the delay period. A physical signature of working memory at work!

Countless experiments since have replicated these findings, showing maintained activity during delay periods under many different circumstances, both in the prefrontal cortex and beyond. Experiments have also hinted that when these firing patterns are out of whack, working memory goes awry. In some experiments, for example, applying a brief electrical stimulation during the delay period can disrupt the ongoing activity and this leads to a dip in performance on delayed response tasks.

What is so special about these neurons that they can do this? Why can they hold on to information and maintain their firing for seconds to minutes, when other neurons let theirs go? For this kind of sustained output, neurons usually need sustained input. But if delay activity occurs without any external input from an image then that sustained input must come from neighbouring neurons. Thus, delay activity can only be generated by a network of neurons working together, the connections between them conspiring to keep the activity alive. This is where the idea of attractors comes back into play.

So far we've looked at attractors in Hopfield networks, which show how input cues reignite a memory. It may not be clear how this helps with working memory. After all, working memory is all about what happens after that ignition; after you stand up to get your roommate's book, how do you keep that goal in mind? As it turns out, however, an attractor is exactly what's needed in this situation, because an attractor stays put.

Attractors are defined by derivatives. If we know the inputs a neuron gets and the weights those inputs are multiplied by, we can write down an equation – a derivative – describing how the activity of that neuron will change over time as a result of those inputs. If this derivative is zero that means there is no change in the activity of the neuron over time; it just keeps firing at the same, constant rate. Recall that, because this neuron is part of a recurrent network, it not only gets input but also serves as input to other neurons. So, its activity goes into calculating the derivative of a neighbouring neuron. If none of the inputs to this neighbouring neuron are changing – that is, their derivatives are all zero as well – then it too will have a zero derivative and will keep firing at the same rate. When a network is in an attractor state, the derivative of each and every neuron in that network is zero.

And that is how, if the connections between neurons are just right, memories started at one point in time can last for much longer. All the cells can maintain their firing rate because all the cells around them are doing the same. Nothing changes if nothing changes.

The problem is that things do change. When you leave the kitchen and walk to your bedroom, you encounter all kinds of things – your shoes in the hallway, the bathroom you meant to clean, the sight of rain on the window – that could cause changes in the input to the neurons that are trying to hold on to the memory. And those changes could push the neurons out of the attractor state representing the book and off to somewhere else entirely. For working memory to function, the network needs to be good at resisting the influence of such distractors. A run-of-the-mill attractor can resist distracting input to an extent. Recall the trampoline example. If the person standing on the trampoline gave a little nudge to the ball, it would likely roll just out of its divot and then back in. With only a small perturbation the memory stays intact, but give the ball a heartier kick and who knows where it will end up? Good memory should be robust to such distractions – so what could make a network good at holding on to memories?

The dance between data and theory is a complex one, with no clear lead or follow. Sometimes mathematical models are developed just to fit a certain dataset. Other times the details from data are absent or ignored and theorists do as their name suggests: theorise about how a system *could* work before knowing how it does. When it comes to building a robust network for working memory, scientists in the 1990s went in the latter direction. They came up with what's known as the 'ring network', a hand-designed model of a neural circuit that would be ideal for the robust maintenance of working memories.

Unlike Hopfield networks, ring networks are well described by their name: they are composed of several neurons arranged in a ring, with each neuron connecting only to those near to it. Like Hopfield networks, these models have attractor states – activity patterns that are self-sustaining and can represent memories. But the attractor states in a ring model are different to those in a Hopfield network. Attractors in a Hopfield model are *discrete*. This means that each attractor state – the one for your childhood bedroom, the one for your childhood holiday, the one for your current bedroom – is entirely isolated from the rest. There is no smooth way to transition between these different memories, regardless of how similar they are; you have to completely leave one attractor state to get to another. Attractors in a ring network, on the other hand, are *continuous*. With continuous attractors, transitioning between similar memories is easy. Rather than being thought of as a trampoline with people standing at different points, models with continuous attractor states are more like the gutter of a bowling lane: once the ball gets into the gutter it can't easily get out, but it can move smoothly within it.

Networks with continuous attractor states like the ring model are helpful for a variety of reasons and chief among them is the type of errors they make. It may seem silly to praise a memory system for its errors – wouldn't we prefer no errors at all? – but if we assume that no network can have perfect memory, then the quality of the errors becomes very important. A ring network allows for small, sensible errors.

Consider the example of the working memory test where subjects had to keep in mind the colour of shapes on a screen. Colours map well to ring networks because, as you'll recall from art class, colours lie on a wheel. So, imagine a network of neurons arranged in a ring, with each neuron representing a slightly different colour. At one side of the ring are red-representing neurons, next to them are orange, then yellow and green; this brings us to the side opposite the red, where there are the blue-representing neurons, which lead to the violet ones and back to red.

In this task, when a shape is seen, it creates activity in the neurons that represent its colour, while the other neurons remain silent. This creates a little 'bump' of activity on the ring, centred on the remembered colour. If any distracting input comes in while the subject tries to hold on to this colour memory – from other random sights in the room, for example – it may push or pull the activity bump away from the desired colour. But – and this is the crucial point – it will only be able to push it to a very nearby place on the ring. So red may become red–orange or green may become teal. But the memory of red would be very unlikely to become green. Or, for that matter, to become no colour at all; that is, there will always be a bump *somewhere* on the ring. These properties are all a direct result of the gutter-like nature of a continuous attractor – it has low resistance for moving between nearby states, but high resistance to perturbations otherwise.

Another benefit of the ring network is that it can be used to do things. The 'working' in working memory is meant to counter the notion that memory is just about passively maintaining information. Rather, holding ideas in working memory lets us combine them with other information and come to new conclusions. An excellent example of this is the head direction system in rats, which also served as the inspiration for early ring network models.

Rats (along with many other animals) have an internal compass: a set of neurons that keep track of the direction the animal is facing at all times. If the animal turns to face a new direction, the activity of these cells changes to reflect that change. Even if the rat sits still in a silent darkened room, these neurons continue to fire, holding on to the information about its direction. In 1995, a team from Bruce McNaughton's lab at the University of Arizona and, separately, Kechen Zhang of the University of California, San Diego, posited that this set of cells could be well described by a ring network. Direction being one of those concepts that maps well to a circle, a bump of activity on the ring would be used to store the direction the animal was facing (See Figure 11).

But not only could a ring network explain how knowledge of head direction was maintained over time, it also served as a model of how the stored direction could change when the animal did. Head direction cells receive input from other neurons, such as those from the visual system and the vestibular system (which keeps track of bodily motion). If these inputs are hooked up to the ring network just right, they can push the bump of activity along to a new place on the ring. If the vestibular system says the body is now moving leftwards, for example, the bump gets pushed to the left. In this way, movement along the ring doesn't create errors in memory, but rather updates the memory based on new information. 'Working' memory earns its name.

Ring networks are a lovely solution to the complex problem of how to create robust and functional working memory systems. They are also beautiful mathematical objects. They display the desirable properties of simplicity and symmetry. They're precise and finely tuned, elegant even.

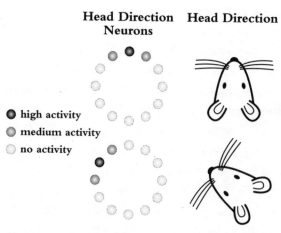

Figure 11

As such, they are completely unrealistic. Because to the biologist, of course, 'finely tuned' are dirty words. Anything that requires delicate planning and pristine conditions to operate well won't survive the chaos that is brain development and activity. Many of the desirable properties of ring networks only occur under very particular assumptions about the connectivity between neurons, assumptions that just don't seem very realistic. So, despite all their desirable theoretical properties and useful abilities, the chances of seeing a ring network in the brain seemed slim.

The discovery made in a research centre just outside Washington DC in 2015 was therefore all the more exciting.

Janelia Research Campus is a world-class research facility hidden away in the idyllic former farmland of Ashburn, Virginia. Vivek Jayaraman has been at Janelia since 2006. He and his team of about half a dozen people work to understand navigation in *Drosophila melanogaster*, a species of fruit fly commonly studied in neuroscience. On a par with a grain of rice, the size of these animals is both a blessing and a curse. While they can be difficult to get hold of, these tiny flies only have around 135,000 neurons, roughly 0.2 per cent the amount of another popular lab animal, the mouse. On top of that, a lot is known about these neurons. Many of them are easily categorised based on the genes they express, and their numbers and locations are very similar across individuals.

Like rodents, flies also have a system for keeping track of head direction. For the fly, these head direction neurons are located in a region known as the ellipsoid body. The ellipsoid body is centrally placed in the fly brain and it has a unique shape: it has a hole in the middle with cells arranged all around that hole, forming a doughnut made of neurons – or in other words, a ring.

Neurons arranged in a ring, however, do not necessarily make a ring network. So, what the Jayaraman lab set out to do was investigate whether this group of neurons that *looked* like a ring network, actually *behaved* like one. To do this, they put a special dye in the ellipsoid body neurons, one that makes them light up green when they're active. They then had the fly walk around, while they filmed the neurons. If you were to look at these neurons on a screen, as the fly heads forwards, you'd see a flickering of little green points at one location on an otherwise black screen. Should the fly choose to make a turn, that flickering patch would swing around to a new location. Over time, as the fly moves and the green patch on the screen moves with it, the points that have lit up form a clear ring structure, matching the underlying shape of the ellipsoid body. If you turn the lights off in the room so the fly has no ability to see which way it's facing, the green flicker still remains at the same location on that ring – a clear sign that the memory of heading direction is being maintained.

In addition to observing the activity on the ring, the experimenters also manipulated it in order to probe the extremes of its behaviour. A true ring network can only support one 'bump' of activity; that is, only neurons at one location on the ring can be active at a given time. So, the researchers artificially stimulated neurons on the side of the ring opposite to those already active. This strong stimulation of the opposing neurons caused the original bump to shut down, and the bump at the new location was maintained, even after the stimulation was turned off. Through these experiments, it became clear that the ellipsoid body was no imposter, but a vivid example of a theory come to life.

This finding – a ring network in the literal, visible shape of a ring – feels a bit like nature winking at us. William Skaggs and the other authors of one of the original papers proposing the ring network explicitly doubted the possibility of such a finding: 'For expository purposes it is helpful to think of the network as a set of circular layers; this does not reflect the anatomical organisation of the corresponding cells in the brain.' Most theorists working on ring network models assumed that they'd be embedded in some larger, messier network of neurons. And that is bound to be the case for most systems in most species. This anomalously pristine example likely arises from a very precisely controlled genetic programme. Others will be much harder to spot.

Even if we usually can't see them directly, we can make predictions about behaviours we'd expect to see if the brain is using continuous attractors. In 1991, pioneering working memory researcher Patricia Goldman-Rakic found that blocking the function of the neuromodulator dopamine made it harder for monkeys to remember the location of items. Dopamine is known to alter the flow of ions into and out of a cell. In 2000, researchers at the Salk Institute in California showed how mimicking the presence of dopamine in a model with a continuous attractor enhanced the model's memory.* It stabilised the activity of the neurons encoding the memory, making

---

* This model was composed of the Hodgkin–Huxley style of neurons described in Chapter 2, which makes incorporating dopamine's effects on ion flows easy.

them more resistant to irrelevant inputs. Because dopamine is associated with reward,* this model also predicts that under conditions where a person is anticipating a large reward their working memory will be better – and that is exactly what has been found. When people are promised more reward in turn for remembering something, their working memory is better. Here, the concept of an attractor works as the thread that stitches chemical changes together with cognitive ones. It links ions to experiences.

Attractors are omnipresent in the physical world. They arise from local interactions between the parts of a system. Whether those parts are atoms in a metal, planets in a solar system or even people in a community, they will be compelled towards an attractor state and, without major disruptions, will stay there. Applying these concepts to the neurons that form a memory connects dots across biology and psychology. On one side, Hopfield networks link the formation and retrieval of memories to the way in which connections between neurons change. On the other, structures like ring networks underlie how ideas are held in the mind. In one simple framework we capture how memories are recorded, retained and reactivated.

---

* Lots more on this in Chapter 11!

CHAPTER FIVE

Excitation and Inhibition

The balanced network and oscillations

Within nearly every neuron, a battle is raging. This fight – a struggle over the ultimate output of the neuron – pits the two fundamental forces of the brain against each other. It's a battle of excitation versus inhibition. Excitatory inputs encourage a neuron to fire. Inhibitory inputs do the opposite: they push the neuron farther from its threshold for spiking.

The balance between these two powers defines the brain's activity. It determines which neurons fire and when. It shapes their rhythms – rhythms that are implicated in everything from attention to sleep to memory. Perhaps more surprisingly, the balance between excitation and inhibition can also explain a feature of the brain that has haunted scientists for decades: the notorious unreliability of neurons.

Eavesdrop on a neuron that should be doing the same thing over and over – for example, one in the motor system that is producing the same movement repeatedly – and you'll find its activity surprisingly irregular. Rather than repeating the same pattern of spikes verbatim each time, it will fire more on some attempts and less on others.

Scientists learned about this peculiar habit of neurons early in the days of neural recordings. In 1932, physiologist Joseph Erlanger made an update to the equipment in his

St Louis laboratory that let him record neural activity with 20 times the sensitivity previously available. He, along with his colleague E. A. Blair, was finally able to isolate individual neurons in the leg of a frog and record how they responded to precise pulses of electricity – 58 identical pulses per minute to be exact.

To their surprise, Erlanger and Blair found that those identical pulses did not produce identical responses: a neuron may respond to one pulse of current, but not the next. There was still a relationship between the strength of the pulse and the response: when weak currents were used, for example, the neuron would respond, say, 10 per cent of the time, medium currents half the time and so on. But beyond these probabilities, how a neuron responded to any given pulse seemed a matter of pure chance. As the pair wrote in their 1933 paper in the *American Journal of Physiology*: 'We were struck by the kaleidoscopic appearance of [responses] obtained from large nerves under absolutely constant conditions.'

This work was one of the first studies to systematically investigate the mysterious irregularity of the nervous system, but it would be far from the last. In 1964, for example, a pair of American scientists applied the same brushing motion to a monkey's skin over and over. They reported that the activity of neurons responding to this motion appeared as 'a series of irregularly recurring impulses, so that, in general, no ordered pattern can be detected by visual inspections'.

In 1983, a group of researchers from Cambridge and New York noted that: 'The variability of cortical neuron response[s] is known to be considerable.' Their study of the visual system in cats and monkeys showed again that the neural response to repeats of the same image produced different results. The responses still had *some* relation to the stimulus – the cells still changed how much they fired *on average* to different images. But exactly which neuron would fire and when for any given instance seemed as unpredictable as next week's weather. 'Successive presentations of identical stimuli do not yield identical responses,' the authors concluded.

In 1998, two prominent neuroscientists even went so far as to liken the workings of the brain to the randomness of radioactive decay, writing that neurons have 'more in common with the ticking of a Geiger counter than of a clock'.

Decades of research and thousands of papers have resulted in a tidy message about just how messy the nervous system is. Signals coming into the brain appear to impinge on neurons already flickering on and off at their own whims. Inputs to these neurons can influence their activity, but not control it exactly – there will always be some element of surprise. This presumably useless chatter that distracts from the main message the neuron is trying to send is referred to by neuroscientists as 'noise'.

As Einstein famously said with regard to the new science of quantum mechanics: 'God does not play dice.' So why should the brain? Could there be any good reason for evolution to produce noisy neurons? Some philosophers have claimed that the noise in the brain could be a source of our free will – a way to overcome a view of the mind as subject to the same deterministic laws of any machine. Yet others disagree. As British philosopher Galen Strawson wrote: 'It may be that some

changes in the way one is are traceable ... to the
influence of indeterministic or random factors. But it is
absurd to suppose that indeterministic or random
factors, for which one is [by definition] in no way
responsible, can in themselves contribute in any way to
one's being truly morally responsible for how one is.' In
other words, following decisions based on a coin flip
isn't exactly 'free' either.

Other purposes for this unpredictability have been
posited by scientists. Randomness, for example, can help
to learn new things. If someone walks the same path to
work every day, occasionally taking a random left turn
could expose them to an unknown park, a new coffee
shop or even a faster path. Neurons might benefit from a
little exploration as well and noise lets them do that.

In addition to the question of *why* neurons are noisy,
the question of *how* they end up this way has preoccupied
neuroscientists. Possible sources of noise exist outside
the brain. Photoreceptors in the eye, for example, need
to be hit by a certain number of photons before they
respond. But even a constant source of light can't
guarantee that a constant stream of photons will reach
the eye. In this way, the input to the nervous system
itself could be unreliable.

In addition, several elements of a neuron's function
depend on random processes. The electrical state of a
neuron, for example, will change if the diffusion of ions
in the fluid around it does. And neurons, like any other
cells, are made up of molecular machines that don't
always function according to plan: necessary proteins
might not be produced fast enough, moving parts may
get stuck, *etc*. While these physical failures could

contribute to the brain's noisiness they don't seem to fully account for it. In fact, when neurons are taken from the cortex and put into a Petri dish they behave remarkably more reliably: stimulating these neurons the same way twice will actually produce similar results. Therefore, basic lapses in cellular machinery – which can happen in the dish just as well as in the brain – seem insufficient to explain the noise that is normally observed.

The books are therefore not balanced: the noise put in somehow doesn't equal the noise produced. We could suspect this is just a curious error in accounting; perhaps there are a few extra unreliable cogs in the neural machinery, or inputs from the world are even less stable than we believe. Such mis-estimates could maybe make up the difference, if it weren't for one small fact: the very nature of how neurons work makes them noise *reducers*.

To understand this, imagine you and some friends are playing a game where the goal is simply to see how far you can collectively move a football down a long field before a timer runs out. None of you are very well practised and occasionally you make mistakes – one person misses a pass, another gets tired, someone else trips. You also occasionally exceed your own expectations by running extra fast or passing extra far. If the time allotted is small – say 30 seconds – such momentary lapses or advantages will have big effects on your distance. You could go 150m on one try and 20 the next. But if the time is large, say five minutes, these fluctuations in performance may simply balance each other out: a slow start could be made up for with an intense sprint at the end, or the gain from a long pass could be lost because of a fall. As a result, the longer the time, the more similar the distance will be on each try. In other words, the 'noisiness' of your athletic ability gets averaged out over time.

Neurons find themselves in a similar situation. If a neuron gets enough input within a certain amount of time, it will fire a spike (see Figure 12). The input it gets is noisy because it comes from the firing of other neurons. So, the neuron may receive, say, five inputs at one moment, 13 the next and zero after that. Just like in the game example, if the neuron takes in this noisy input for a long time before deciding if it has enough to spike, the impact of the noise will be reduced. If it only uses a quick snapshot of its input, however, the noise will dominate.

So how much time does a neuron combine its inputs over? About 20 milliseconds. That may not seem like long, but for a neuron it's plenty. A spike only takes about 1 millisecond and a cell can be receiving many at a time from all of its different inputs. Therefore, neurons should be able to take the average over many snapshots of input before deciding to spike.

Neuroscientists William Softky and Christof Koch used a simple mathematical model of a neuron – the 'leaky integrate-and-fire' model introduced in Chapter 2 – to test just this. In their 1993 study, they simulated a neuron receiving inputs at irregular times. Yet the neuron itself – because it integrated these incoming spikes over time – still produced output spikes that were much more regular than the input it received. This means neurons do have the power to destroy noise – to take in noisy inputs while producing less noisy outputs.

If neurons weren't able to quench noise, then the unreliability of the brain wouldn't be as much of a

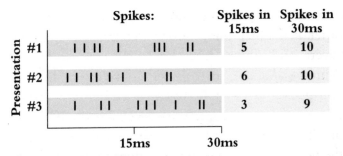

Over three different presentations of the same stimulus the neuron fires spikes at very different times. If another neuron got input from this one, but only listened to a short snippet of its spikes, it would receive a different number of spikes on each presentation. Over a longer time, the number is more similar.

Figure 12

mystery. As mentioned before, we could assume that small amounts of randomness enter the brain – either from the outside world or from inside a cell – and spread through the system via the connections between neurons. If noisy inputs led to outputs that were just as, or possibly more, noisy, this would be a perfectly self-consistent story; noisy neurons would beget noisy neurons. But, according to Softky and Koch's model, this is not what happens. When passed through a neuron, noise should get weaker. When passed over and over through a whole network of neurons, it should be expected to all but disappear. Yet everywhere neuroscientists look, there it is.

Not only is the brain unpredictable, then, but it seems to be encouraging that unpredictability – going against the natural tendency of neurons to squash it. What's keeping the randomness alive? Does the brain have a random-number generator? Some sort of hidden biological dice? Or, as scientists in the 1990s hypothesised, does all this disorder actually result from a more fundamental order, from the balance between excitation and inhibition?

★ ★ ★

It took Ernst Florey several trips to a Los Angeles horse butcher to uncover the source of inhibition in the brain.

It was the mid-1950s when Florey, a German-born neurobiologist who emigrated to North America, was working on this question with his wife Elisabeth. At that time, the fact that neurons communicate by sending chemicals – called neurotransmitters – between each other was largely established. However, the only known neurotransmitters were excitatory – that is, they were chemicals that made a neuron *more* likely to fire. Yet since the mid-nineteenth century, it was known that some neurons can actually reduce the electrical activity of their targets. For example, the Weber brothers, Ernst and Eduard, showed in 1845 that electrically stimulating a nerve in the spinal cord could slow down the cells that controlled the beating of the heart, even bringing it to a standstill. This meant the chemical released by these neurons was inhibitory – it made cells less likely to fire.

Florey needed specimens for his study on 'factor I', his name for the substance responsible for inhibition. So, he regularly rode his 1934 Chevrolet to a horse butcher and picked up some parts less favoured by their average customers: fresh brains and spinal cords. After extracting different substances from this nervous tissue, he checked what happened when each one was applied to living neurons taken from a crayfish. Eventually he identified some candidate chemicals that reliably quieted the crayfish neurons. This cross-species pairing was actually a bit of luck on Florey's part. Neurotransmitters can't always be assumed to function the same way across animals. But in this case, what was inhibitory for the horse was inhibitory for the crayfish.

With the help of professional chemists, Florey then used tissue from yet another animal – 45kg (100lb) of cow brain to be precise – to purify 'factor I' down to its most basic chemical structure. In the end he was left with 18mg of gamma-aminobutyric acid. Gamma-aminobutyric acid (or GABA, as it is more commonly known) was the first identified inhibitory neurotransmitter.

Whether a neurotransmitter is inhibitory or excitatory is really in the eye of the beholder – or more technically in the receptor of the target neuron. When a neurotransmitter is released from one neuron, the chemical travels the short distance across the synapse between that neuron and its target. It then attaches itself to receptors that line the membrane of the target neuron. These receptors are like little protein padlocks. They require the right key – that is, the correct neurotransmitter – to open. And once they open they're pretty selective about who they let in. One kind of receptor that GABA attaches to, for example, only lets chloride ions into the cell. Chloride ions have a negative charge and letting more in makes it harder for the neuron to reach the electrical threshold it needs to fire. The receptors to which excitatory neurotransmitters attach let in positively charged ions, like sodium, which bring the neuron *closer* to threshold.

Neurons tend to release the same neurotransmitter on to all of their targets, a principle known as Dale's Law (named after Henry Hallett Dale who boldly guessed as much in 1934, a time at which only two neurotransmitters had even been identified). Neurons that release GABA are called 'GABAergic', although because GABA is the most prevalent inhibitory neurotransmitter in the adult mammalian brain, they're frequently just called 'inhibitory'. Excitatory transmitters are a bit more diverse, but the neurons that release them are still broadly classed as 'excitatory'. Within an area of the cortex, excitatory and inhibitory neurons freely intermix, sending connections to, and receiving them from, each other.

In 1991, after many of these facts of inhibition had been established, Florey wrote a retrospective on his role in the discovery of the first – and arguably most important – inhibitory neurotransmitter. He ended it with the sentence: 'Whatever the brain does for the mind, we can be sure that GABA plays a major role in it.' Likely unbeknownst to Florey, at the same time a theory that makes inhibition a key player in the production of the brain's unpredictability was developing.

★ ★ ★

Returning to the analogy of the timed football game, imagine now that another team is added. Their goal is to fight against you, moving the ball to the opposite end of the field. When the clock stops, whoever is closer to their goal side wins. If the other team is also made up of your semi-athletic friends, then on average both teams would perform the same. The noisiness of your performance will still affect the outcome: your team may beat the other by a few metres on one try and get beaten by the same small amount on another. But on the whole, it will be a balanced and tame game.

Now consider if the other team was made up of professional athletes – some of the strongest, fastest players around. In this case, you and your friends wouldn't stand a chance; you'd be clobbered every time. This is why no one would bother watching a competition between career football players and the high school team, Tiger Woods vs your dad, or Godzilla against a literal moth. The outcome of all those matches is just too predictable to be interesting. In other words, unfair fights create consistency; fair fights are more fun.

In the cortex, neurons get thousands of connections from both excitatory and inhibitory cells. Because of this, each individual force is strong and would consistently dominate if the other were any weaker. Without the presence of inhibition, for example, the hundreds of excitatory inputs bombarding a cell at any moment would make it fire almost constantly; on the other hand, inhibition alone would drive the cell down to a completely stagnant state. With huge power on each side, the true activity of the neuron is thus the result of a tug of war between giants. What's happening in the neuron is indeed a balanced fight and it's the kind you'd see at the Olympics rather than in a schoolyard.

Tell this fact to a computer scientist and they may start to get worried. That's because computer scientists know that taking the difference between two big, noisy numbers can cause big problems. In computers, numbers can only be represented with a certain level of precision. This means that some numbers need to be rounded off, introducing error – or noise – into the computation. For example, a computer with only three digits of precision may represent the number 18,231 as 1.82×10^4; the leftover 31 is lost in the rounding error. When subtracting two roughly equal numbers, the effects of this rounding error can dominate the answer. For example, 18,231 minus 18,115 is 116, yet the computer would calculate this difference as 1.82×10^4 minus 1.81×10^4 which is only 100! That puts the computer off by 16. And the larger the numbers, the larger the error will be. For example, a computer with three-digit precision calculating 182,310 minus 181,150 would produce an answer that is 160 less than the true one.

You reasonably wouldn't feel comfortable if your bank or doctor's office were doing their calculations this way. For this reason, programmers are taught to write their code in a way that avoids subtracting two very large numbers. Yet neurons are subtracting two large numbers – excitation minus inhibition – at every moment. Could such a 'bug' really be part of the brain's operating system?

Scientists had been toying with this idea for a while when, in 1994, Stanford neuroscientists Michael Shadlen and William Newsome decided to test it out. Similar to the work of Softky and Koch, Shadlen and Newsome built a mathematical model of a single neuron and fed it inputs. This time, however, the neuron got both noisy excitatory *and* noisy inhibitory inputs. When pitting these two forces against each other, sometimes excitation will win and sometimes inhibition will. Would the fight play out like a noisy calculation and create a neuron that fired erratically? Or would the neuron still be able to crush the noise in these inputs the way it had the excitatory inputs in Softky and Koch's work? Shadlen and Newsome found that, indeed, given both these types of inputs – each coming in at the same, high rate – the neuron's output was noisy.

In a boxing match between amateurs, a momentary lapse in attention from one may let the other land a small hit. In a fight between pros, however, that same lapse could lead to a knockout. In general, the stronger that two competing powers are, the bigger the swings in the outcome of their competition. This is how the internal struggle between excitation and inhibition in a neuron can overpower its normal noise-crushing abilities. Because both sources are evenly matched, the neuron's net input (that is the total excitation minus the total inhibition) is not very big on average. But because both sources are strong, the swings around that average are huge. At one moment the neuron can get pushed far above its threshold for firing and emit a spike. At the next it could be forced into silence by a wave of inhibition. These influences can make a neuron fire when it otherwise wouldn't have or stay silent when it otherwise would have. In this way, a balance between excitation and inhibition creates havoc in a neuron and helps explain the variability of the brain.

The simulation run by Shadlen and Newsome went a long way in helping to understand how neurons could remain noisy. But it didn't quite go far enough. Real neurons get inputs from other real neurons. For the theory that noise results from a balance between excitation and inhibition to be correct, it thus has to work for a whole network of excitatory and inhibitory neurons. That means one in which each neuron's input comes from the other neurons and its outputs go back to them too. Shadlen and Newsome's simulation, however, was just of a single neuron, one that received inputs controlled by the model-makers. You can't just look at the income and expenses of a single household and decide that the national economy is strong. Similarly, simulating a single neuron can't guarantee that a network of neurons will work as needed. As we saw in the last chapter, in a system with a lot of moving parts, all of them need to move just right to get the desired outcome.

To get a whole network to produce reliably noisy neurons requires coordination: each neuron needs to get excitatory and inhibitory input from its neighbours in roughly equal proportions. And the network needs to be *self-consistent* – that is, each neuron needs to produce the same amount of noise it receives, no more nor less. Could a network of interacting excitatory and inhibitory cells actually sustain the kind of noisy firing seen in the brain, or would the noise eventually peter out or explode?

★ ★ ★

When it comes to questions of self-consistency in networks, physicists know what to do. As we saw in the last chapter, physics is full of situations where self-consistency is important: gases made of large numbers of simple particles, for example, where each particle is influenced by all those around it and influences them all in return. So, techniques have been developed to make the maths of these interactions easier to work with.*

In the 1980s, Israeli physicist Haim Sompolinsky was using these techniques to understand the ways materials behave at different temperatures. But his interests eventually turned towards neurons. In 1996, Sompolinsky and fellow physicist-turned-neuroscientist Carl van Vreeswijk, applied the physicist's approach to the question of balance in the brain. Mimicking the mathematics used to understand interacting particles, they wrote down some simple equations that represented a very large population of interacting excitatory and inhibitory cells. The population also received external inputs meant to represent connections coming from other brain areas.

With their simple equations, it was possible for van Vreeswijk and Sompolinsky to mathematically define the kind of behaviour they wanted to see from the model. For example, the cells had to be able to keep themselves active, but not too active (they shouldn't be firing non-stop, for example). In addition, they should respond to increases in external input by increasing their average firing rate. And, of course, the responses had to be noisy.

Putting in these demands, van Vreeswijk and Sompolinsky then churned through the equations. They found that, to create a full network that will keep firing

---

* Historically, this set of techniques went under the more obvious name of 'self-consistent field theory', but it's now known as 'mean-field theory'. The trick behind the mean-field approach is that you don't need to provide an equation for each and every interacting particle in your system. Instead, you can study a 'representative' particle that receives its own output as input. That makes studying self-consistency a lot easier.

away irregularly at a reasonable rate, certain conditions had to be met. For example, the inhibitory cells need to have a stronger influence on the excitatory cells than the excitatory cells have on each other. Ensuring that the excitatory cells receive slightly more inhibition than excitation keeps the network activity in check. It's also important that the connections between neurons be random and rare – each cell should only get inputs from, say, five or ten per cent of the other cells. This ensures no two neurons get locked into the same pattern of behaviour.

None of the requirements van Vreeswijk and Sompolinsky found were unreasonable for a brain to meet. And when the pair ran a simulation of a network that met all of them, the necessary balance between excitation and inhibition emerged, and the simulated neurons looked as noisy as any real ones. Shadlen and Newsome's intuition about how a single neuron can sustain noisy firing did in fact hold in a network of interacting neurons.

Beyond just showing that balancing excitation and inhibition was possible in a network, van Vreeswijk and Sompolinsky also found a possible benefit of it: neurons in a tightly balanced network respond quickly to inputs. When a network is balanced, it's like a driver with each foot pressed equally on the gas and the brake. This balance gets disrupted, however, if the amount of external input changes. Because the external inputs are excitatory – and they target the excitatory cells in the network more than the inhibitory ones – an increase in their firing is like a weight added to the gas pedal. The car then zooms off almost as quickly as the input came in. After this initial *whoosh* of a response, however, the network regains its balance. The explosion of excitation in the network causes the inhibitory neurons to fire more and — like adding an additional weight to the brake — the network resettles in a new equilibrium, ready to respond again. This ability to act so quickly in response to a changing input could help the brain accurately keep up with a changing world.

Knowing that the maths works out is reassuring, but the real test of a theory comes from real neurons. Van Vreeswijk and Sompolinsky's work makes plenty of predictions for neuroscientists to test, and that's just what Michael Wehr and Anthony Zador at Cold Spring Harbor Laboratory did in 2003. The pair recorded from neurons in the auditory cortex of rats, which is responsible for processing sound, while different sounds were played to the animal. Normally when neuroscientists drop an electrode into the brain they're trying to pick up on the output of neurons — that is, their spikes. But these researchers used a different technique to eavesdrop on the input a neuron was getting — specifically to see if the excitatory and inhibitory inputs balanced each other out.

What they saw was that, right after the sound turned on, a strong surge of excitation came into the cell. It was followed almost immediately by an equal influx of inhibition — the brake that follows the gas. Therefore, increasing the input to this real network showed just the behaviour expected from the model. Even when using louder sounds that produced more excitation, the amount of inhibition that followed always matched it. Balance seemed to be emerging in the brain just as it did in the model.

To explore another prediction of the model, scientists had to get a bit creative. Van Vreeswijk and Sompolinsky showed that, to make a well-balanced network, the strength of connections between neurons should depend on the total number of connections: with more connections, each connection can be weaker. Jérémie Barral and Alex Reyes from New York University wanted a way to change the number of connections in a network in order to test this hypothesis.

Within a brain it's hard to control just how neurons grow. So, in 2016, they decided to grow them in a Petri dish instead. It's an experimental set-up that – in its simplicity, controllability and flexibility – is almost like a live version of a computer simulation. In order to control the number of connections, they simply put different numbers of neurons in the dish; dishes with more neurons packed in made more connections. They then monitored the activity of the neurons and checked their connection strengths. All the populations (which contained both excitatory and inhibitory cells) fired noisily, just as a balanced network should. But the connection strengths varied drastically. In a dish where each neuron only got about 50 connections, the connections were three times stronger than those with 500 connections. In fact, looking across all the populations, the average strength of a connection was roughly equal to one divided by the square root of the number of connections – exactly what van Vreeswijk and Sompolinsky's theory predicted.

As more and more evidence was sought, more was found for the belief that the brain was in a balanced state. But not all experiments went as the theory predicted; tight balance between excitation and inhibition wasn't always seen. There is good reason to believe that certain brain areas engaged in certain tasks may be more likely to exhibit balanced behaviour. The auditory cortex, for example, needs to respond to quick changes in sound frequency to process incoming information. This makes the quick responsiveness of well-balanced neurons a good match. Other areas that don't require such speed may find a different solution.

The beauty of balance is that it takes a ubiquitous inhabitant of the brain – inhibition – and puts it to work solving an equally ubiquitous mystery – noise. And it does it all without any reliance on magic: that is, no hidden source of randomness. The noise comes even while neurons are responding just as they should.

This counter-intuitive fact that good behaviour can produce bedlam is important. And it had been observed somewhere else before. Van Vreeswijk and Sompolinsky make reference to this history in the first word of the title of their paper: 'Chaos in neuronal networks with balanced excitatory and inhibitory activity'.

★ ★ ★

Chaos didn't exist in the 1930s: when neuroscientists were first realising how noisy neurons are, the mathematical theory to understand their behaviour hadn't yet been discovered. When it was, it happened seemingly by chance.

The Department of Meteorology at MIT was founded in 1941, just in time for Edward Lorenz. Lorenz, born in 1917 in a nice Connecticut neighbourhood to an engineer and a teacher, had an early interest in numbers, maps and the planets. He intended to continue on in mathematics after earning an undergraduate degree in it, but, as was the case for so many scientists of his time, the war intervened. In 1942 Lorenz was given the task of weather prediction for the US Army Air Corps. To learn how to do this, he took a crash course in meteorology at MIT. When he was done with the army he stayed with meteorology and remained at MIT: first as a PhD student, then a research scientist and finally a professor.

If you've ever tried to plan a picnic, you know weather prediction is far from perfect. Academic meteorologists, who focus on the large-scale physics of the planet, hardly even consider day-to-day forecasting a goal. But Lorenz remained curious about it and about how a new technology – the computer – could help.

The equations that describe the weather are many and complex. Churning through them by hand – to see how the weather right now will lead to the weather later – is an enormous, nearly impossible task (by the time you finish it the weather you're predicting likely will have passed). But a computer could probably do it much faster.

Starting in 1958, Lorenz put this to the test. He boiled the dynamics of weather down to 12 equations, picked some values to start them off with – say, westerly winds at 100km/hr – and let the mathematics run. He printed the output of the model on rolls of paper as it went along. It looked weather-like enough, with familiar ebbs and flows of currents and temperatures. One day he wanted to re-run a particular simulation to see how it would evolve over a longer period of time. Rather than starting it from the very beginning again, he figured he could start it part way along by putting in the values from the printout as the starting conditions. Impatience, sometimes, is the mother of discovery.

The numbers that the computer printed out, however, weren't the full thing. To fit more on the page, the printer cut the number of digits after the decimal point from six down to three. So, the numbers Lorenz put in for the second run of the simulation weren't *exactly* where the model was before. But what could a few decimal points matter in a model of the whole world's weather? Turns out quite a bit. After a few cranks of the mathematical machinery – about two months of weather changes in model time – this second run of the simulation was completely different from the first. What was hot was cold, what was fast was slow. What was supposed to be a replication turned into a revelation.

Up until that point, scientists assumed that small changes only beget small changes. A little gust of wind at one point in time should have no power to move mountains later. Under that dogma, what Lorenz observed must've come from a mistake, a technical error made by the large, clunky computers of the day, perhaps.

Lorenz was willing to see what was truly happening, however. As he wrote in 1991: 'The scientist must always be on the lookout for other explanations than those that have been commonly disseminated.' What Lorenz had observed was the true behaviour of the mathematics, as counter-intuitive as it seemed. In certain situations, small fluctuations *can* get amplified, making behaviour unpredictable. It's not a mistake or an error; it's just how complex systems work. Chaos – the name given to this phenomenon by mathematicians* – was real and it would do scientists well to try to understand it.

Chaotic processes produce outputs that *look* random but in fact arise from perfect rule-following. The source of this deception is the upsetting truth that our ability to predict outcomes based on knowing the rules is far more limited than previously thought – especially if those rules are complex. In his book *Chaos: Making a New Science*, a sweeping history of how the field emerged, James Gleick wrote: 'Traditionally, a dynamicist would believe that to write down a system's equations is to understand the system ... But because of the little bits of nonlinearity in these equations, a dynamicist would find himself helpless to answer the easiest practical questions about the future of the system.' This imbues even the simplest systems of, say, interacting billiard balls or swinging pendulums with the potential to produce something surprising. He continued: 'Those studying chaotic dynamics discovered that the disorderly behaviour of simple systems acted as a *creative* process. It generated complexity: richly organised patterns, sometimes stable and sometimes unstable, sometimes finite and sometimes infinite.'

Chaos was happening in the atmosphere – and if van Vreeswijk and Sompolinsky were right, it was happening in the brain, too. For this reason, explaining why the brain reacts to repeated inputs with a

---

* In popular culture it's better known as the 'butterfly effect', the idea that something as insignificant as a butterfly flapping its wings can change the whole course of history.

kaleidoscopic variety needn't involve spotty cellular machinery. That's not to say that there aren't any sources of noise in the brain (such as unreliable ion channels or broken-down receptors), but just that an object as complex as the brain, with its interacting pools of excitation and inhibition, doesn't require them to show rich and irregular responses. In fact, in their simulation of a network, all van Vreeswijk and Sompolinsky had to do was change the starting state of a single neuron – from firing to not, or vice versa – to create a completely different pattern of activity across the population.* If a change so small can create such a disturbance, the brain's ability to keep noise alive seems less mysterious.

★ ★ ★

In medical centres around the world, epilepsy patients spend several days – up to a week – stuck in small rooms. These 'monitoring' rooms are usually equipped with a TV – for the patients – and cameras that monitor patient movement – for the doctors. All day and night, the patients are connected to an electroencephalogram (EEG) machine that's capturing their brain's behaviour. They hope that the information gathered will help to treat their seizures.

---

* Again, the population will still produce the same amount of spikes *on average* in response to a given input. It's just how those spikes are distributed across time and neurons that varies. If your neurons were truly following no rules for how they responded to inputs, you wouldn't be able to be reading this right now.

EEG electrodes, attached via stickers and tape to the scalp, monitor the electrical activity produced by the brain below. Each electrode provides one measurement – a complicated combination of the activity of many, many neurons at once. It's a signal that varies over time like a seismograph. When patients are awake, the signal is a jagged and squiggly line: it moves slightly up and slightly down at random, but without any strong rhythm. When patients are asleep (particularly in deep dreamless sleep), the EEG makes waves: large movements upwards then downwards extending over a second or more. When the event of interest – a seizure – occurs, the movements are even starker. The signal traces out big, fast sweeps up and down, three to four times a second, like a kid scribbling frantically with a crayon.

What are neurons doing to create these strong signals during a seizure? They're working together. Like a well-trained military formation, they march in lockstep: firing in unison then remaining silent before firing again. The result is a repeated, synchronous burst of activity that drives the EEG signal up and down over and over again. In this way, a seizure is the opposite of randomness – it is perfect order and predictability.

The same neurons that produce that seizure also produce the slow waves of sleep and the normal, noisy activity needed for everyday cognition. How can the same circuit exhibit these different behaviours? And how does it switch between them?

In the late 1990s, French computational neuroscientist Nicolas Brunel set out to understand the different ways circuits can conduct themselves.* Specifically, building off the work of van Vreeswijk and Sompolinsky, he wanted to investigate how models made of excitatory and inhibitory neurons behave. To do this, Brunel explored the *parameter space* of these models.

Parameters are the knobs that can be turned on a model. They are values that define specific features, like the number of neurons in the network or how many inputs each gets. Like regular space, parameter space can be explored in many different directions, but here each direction corresponds to a different parameter. The two parameters Brunel chose to explore were, firstly, how much external input the network gets (*i.e.*, input from other brain areas) and, secondly, how strong the inhibitory connections in the network are compared with the excitatory ones. By changing each of these parameters a bit and churning through the equations, Brunel could check how the behaviour of the network depends on these values.

Doing this for a bunch of different parameter values results in a map of the model's behaviour. The latitude and longitude on this map (see Figure 13) correspond to the two parameters Brunel varied respectively. For the network at the middle of the map, the inhibition is exactly equal to the excitation and the input to the network is of medium strength. Moving to the left on the map, excitation becomes stronger than inhibition;

---

* Brunel, perhaps unsurprisingly at this point, started as a physicist. He learned about neuroscience during his PhD in the early 1990s, when a course exposed him to this new trend of applying the tools of physics to the brain.

move to the right and vice versa. Move upwards and the input to the network gets stronger, down and it's weaker. Laid out this way, the network van Vreeswijk and Sompolinsky studied – with the inhibitory connections slightly stronger than excitatory ones – is just off to the right of the middle.

Brunel surveyed this model landscape looking for any changes in the terrain: do certain sets of parameters make the network behave drastically differently? To find the first striking landmark you don't have to travel far from van Vreeswijk and Sompolinsky's original network. Crossing over from the region where inhibition is stronger into the one where excitation is, a sharp transition happens. In mathematics, these

The Map

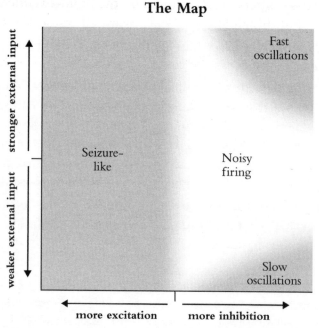

Figure 13

transitions are known as bifurcations. Like a steep cliff separating a grassy plain from a sea, bifurcations mark a quick change between two separate areas in parameter space. In Brunel's map, the line where excitation and inhibition are equal separates the networks with irregular, noisy firing on the right from those with rigid, predictable firing on the left. Specifically, when inhibition gets too weak, the neurons in these networks stop their unique pitter-patter and start firing in unison. Their tight synchronous activity – with groups of neurons flicking on and off together – looks a lot like a seizure.

Physiologists have known for centuries that certain substances act as convulsants – that is, they induce seizures. With the increased understanding of neurotransmitters that came in the mid-twentieth century, it became clear that many of these drugs interfered with inhibition. Bicuculline, for example, is found in plants across North America and stops GABA from attaching to its receptor. Thujone, present in low doses in absinthe, prevents GABA receptors from letting chloride ions in. Whatever the mechanism, in the end, these drugs are throwing the balance off in the brain, putting inhibitory influences at a disadvantage. Using his bird's-eye view of the brain's behaviour, Brunel could see how changing the brain's parameters – through drugs or otherwise – moved it into different states.

Travelling to the other end of Brunel's map reveals yet another pattern of activity. In this realm, inhibition rules over excitation. If the external input remains at medium strength, the neurons remain noisy here. Move up or down, however, and two similar, but different, behaviours appear. With both high and low external input, the neurons show some cohesion. If you added up how many neurons were active at any given time, you'd see waves of activity: brief periods of more-than-average firing followed by less. But unlike the military precision of the seizure state, networks here are more like a percussion section made up of six-year-olds: there may be some organisation, but not everybody is playing together all the time. In fact, an individual neuron in these networks only participates in every third or fourth wave – and even then their timing isn't always perfect. In this way, these states are both oscillating and noisy.

The feature that differentiates behaviour in the top right corner of the map from that in the bottom right is the *frequency* of that oscillation. Drive the network with strong external inputs and its average activity will move up and down quickly – as fast as 180 times per second. The strong input drives the excitatory cells that drive the inhibitory cells to shut them down; then the inhibitory cells shut themselves down and the whole thing repeats. Reduce the network's input and it oscillates more slowly, around 20 times a second. These slow oscillations occur because the external input to the network is so weak, and inhibition is so strong, that many neurons just don't get enough input to fire. The ones that are firing, however, use their connections to slowly rev the network back up. If too many inhibitory cells get activated, though, the network becomes quiet again.

Despite their superficial similarity to a seizure, these messy oscillations aren't actually debilitating. In fact, scientists have observed oscillations in all different parts of the brain under all different conditions. Groups of neurons in the visual cortex, for example, can oscillate at a swift 60 times a second. The hippocampus (the memory-processing machine from last chapter) sometimes oscillates quickly and other times slowly. The olfactory bulb, where odours are processed, generates waves that range from once per second – aligning with inhalation – to a hundred times. Oscillations can be found everywhere if you care to look.

Mathematicians are happy to see oscillations. This is because, to a mathematician, oscillations are approachable. Chaos and randomness are a challenge to capture with equations, but perfect periodicity is easy and it's elegant. Over millennia, mathematicians have developed the equipment not just to describe oscillations, but to predict how they'll interact and to spot them in signals that – to the untrained eye – may not look like oscillations at all.

Nancy Kopell is a mathematician, or at least she used to be. Like her mother and sister before her, Kopell majored in mathematics as an undergraduate. She then got a PhD* in it from the University of California, Berkeley, in 1967, and became a professor of mathematics

---

* Kopell's reasons for going to graduate school were somewhat unusual: 'I had not entered college thinking about going to graduate school. But when my senior year arrived, I was not married and had nothing specific I wanted to do, so graduate school seemed like a good option.' But the sexism she encountered there was perhaps more expected: 'It was the unspoken but very widespread assumption that women in math were like dancing bears – perhaps they could do it, but not very well, and the attempt was an amusing spectacle.'

at Northeastern University in Boston. But after many years crossing back and forth between the border of mathematics and biology – taking problems from the latter to inspire ideas for the former – she started to feel more settled in the land of biology. As Kopell wrote in an autobiography: 'My perspective started to shift, and I found myself at least as interested in the physiological phenomena as the mathematics problems that they generated. I didn't stop thinking mathematically, but problems interested me less if I didn't see the relevance to specific biological networks.' Many of the biological networks that interested her were neural ones and throughout her career she's studied all manner of oscillations in the brain.

High-frequency oscillations are referred to by neuroscientists as 'gamma' waves. The reason for this is that Hans Berger, the inventor of the original EEG machine, called the big slow waves he could see by eye on his shoddy equipment 'alpha' waves and everything else 'beta'; the scientists who came after him simply followed suit, giving new frequencies they found new Greek letters. Gamma waves, while fast, are usually small – or 'low amplitude' in technical terms. Their presence, detectable by a modern EEG or an electrode in the brain, is associated with an alert and attentive mind.

In 2005, Kopell and colleagues came up with an explanation for how gamma oscillations could help the brain focus. Their theory stems from the idea that neurons representing the information you're paying attention to should get a head start in the oscillation. Consider trying to listen to a phone call in the middle of a noisy room. Here, the signal you are paying attention to – the voice on the other end of the line – is competing against all the distracting sounds in the room. In Kopell's model, the voice is represented by one group of excitatory cells and the background chatter by another. Both of these groups send connections to a common pool of inhibitory neurons and both get connections back from it in return.

Importantly, the neurons representing the voice – because they are the object of attention – get a little more input than the 'background' neurons. This means they'll fire first and more vigorously. If these 'voice' neurons fire in unison, they will – through their connections to the inhibitory cells – cause a big, sharp increase in inhibitory cell-firing. This wave of inhibition will then shut down the cells representing both the voice and the background noise. Because of this, the background neurons never get the chance to fire and therefore can't interfere with the sound of the voice. It's as though the voice neurons, by being the first to fire, are pushing themselves through a door and then slamming it shut on the background neurons. And as long as the voice neurons keep getting a little extra input, this process will repeat over and over – creating an oscillation. The background neurons will be forced to remain silent each time. This leaves just the clean sound of the voice in the phone as the only remaining signal.

Beyond just this role in attention, neuroscientists have devised countless other ways in which oscillations could help the brain. These include uses in navigation, memory and movement. Oscillations are also supposed to make communication between brain areas better and help organise neurons into separately functioning groups. On top of this, theories abound for how oscillations go wrong in diseases like schizophrenia, bipolar disorder and autism.

The omnipresence of oscillations may make it seem like their importance would go unquestioned, but this is far from the case. While several different roles for oscillations have been devised, many scientists remain sceptical.

Part of the concern comes from the very first step: how oscillations are measured. Instead of recording from many neurons at once, many researchers interested in oscillations use an indirect measure that comes from the fluid that surrounds neurons. Specifically, when neurons are getting a lot of input, the composition of ions in this fluid changes and this can be used as a proxy for how active the population is. But the relationship between ion flows in this fluid and the real activity of neurons is complicated and not completely understood. This makes it hard to know if observed oscillations are actually happening.

Scientists can also be swayed by the tools available to them. EEG has been around for a century and it makes the spotting of oscillations easy, even in human subjects; experiments can be performed in an afternoon on willing (usually undergraduate) research participants. As mentioned, a similar ease and familiarity applies to the mathematical tools of analysing oscillations. This may make researchers more likely to seek out these brainwaves, even in cases where they might not provide the best answers. To paraphrase the old adage, when a hammer is the easiest tool you have to use, you start looking for nails.

Another issue is impact, especially when it comes to fast oscillations like gamma. If one brain state has stronger gamma waves than another, it means more neurons are firing as part of a wave in that state rather than sporadically on their own. But when these waves come so quickly, being part of one may only make a neuron fire a few milliseconds before or after it otherwise would have. Could that kind of temporal precision really matter? Or is all that matters the total number of spikes produced? Many elegant hypotheses about how oscillations can help haven't been directly tested – and can be quite hard to test – so answers are unknown.

As neuroscientist Chris Moore said in a 2019 *Science Daily* interview: 'Gamma rhythms have been a huge topic of debate … Some greatly respected neuroscientists view gamma rhythms as the magic, unifying clock that aligns signals across brain areas. There are other equally respected neuroscientists that, colourfully, view gamma rhythms as the exhaust fumes of computation: They show up when the engine is running but they're absolutely not important.'

Exhaust fumes may be produced when a car moves, but they're not directly what's making it go. Similarly, networks of neurons may produce oscillations when performing computations, but whether those oscillations are what's doing the computing remains to be seen.

As has been shown, the interaction between excitatory and inhibitory cells can create a zoo of different firing patterns. Putting these two forces in conflict has both benefits and risks. It grants the network the ability to respond with lightning speed and to generate the smooth rhythms needed for sleep. At the same time, it places the brain dangerously close to seizures and creates literal chaos. Making sense of such a multifaceted system can be a challenge. Luckily a diversity of mathematical methods – those developed for physics, meteorology and understanding oscillations – have helped tame the wild nature of neural firing.

CHAPTER SIX

Stages of Sight

The Neocognitron and convolutional neural networks

The summer vision project is an attempt to use our summer workers effectively in the construction of a significant part of a visual system. The particular task was chosen partly because it can be segmented into sub-problems which will allow individuals to work independently and yet participate in the construction of a system complex enough to be a real landmark in the development of 'pattern recognition'.

Vision Memo No. 100 from the
Massachusetts Institute of Technology
Artificial Intelligence Group, 1966

The summer of 1966 was meant to be the summer that a group of MIT professors solved the problem of artificial vision. The 'summer workers' they planned to use so effectively for this project were a group of a dozen or so undergraduate students at the university. In their memo laying out the project's plan, the professors provided several specific skills they wanted the computer system the students were developing to perform. It should be able to define different textures and lighting in an image, label parts as foreground and parts as background, and identify whatever objects were present. One professor* purportedly described the aims more casually as 'linking a camera to a computer and getting the computer to describe what it saw'.

The goals of this project were not completed that summer. Nor the next. Nor many after that. Indeed, some of the core issues raised in the description of the summer project remain open problems to this day. The hubris on display in that memo is not surprising for its time. As discussed in Chapter 3, the 1960s saw an explosion in computing abilities and, in turn, naive hopes about automating even the most complex tasks. If computers could now do anything asked of them, it was just a matter of knowing what to ask for. With something as simple and immediate as vision, how hard could that be?

The answer is very hard. The act of visual processing – of taking in light through our eyes and making sense of the external world that reflected it – is an immensely complex one. Common sayings like 'right in front of your eyes' or 'in plain sight', which are used to indicate the effortlessness of vision, are deceitful. They obscure the significant challenges even the most basic visual inputs pose for the brain. Any sense of ease we have regarding vision is an illusion, one that was hard won through millions of years of evolution.

The problem of vision is specifically one of reverse engineering. In the back of the eye, in the retina, there is

---

* That professor would be Marvin Minsky and the professor that wrote the memo was Seymour Papert, both key participants in Chapter 3. Indeed, as you'll see, there are many overlapping players and themes in the histories of artificial neural networks and artificial vision.

a wide flat sheet of cells called photoreceptors. These cells are sensitive to light. Each one indicates the presence or absence (and possibly wavelength) of the light hitting it at each moment by sending off a signal in the form of electrical activity. This two-dimensional flickering map of cellular activity is the only information from which the brain is allowed to reconstruct the three-dimensional world in front of it.

Even something as simple as finding a chair in a room is a technically daunting endeavour. Chairs can be many different shapes and colours. They can also be nearby or far away, which makes their reflection on the retina larger or smaller. Is it bright in the room or dark? What direction is light coming from? Is the chair turned towards you or away? All of these factors impact the exact way in which photons of light hit the retina. But trillions of different patterns of light could end up meaning the same thing: a chair is there. The visual system somehow finds a way to solve this many-to-one mapping in less than a tenth of a second.

At the time those MIT students were working to give the gift of sight to computers, physiologists were using their own tools to solve the mysteries of vision. This started with recording neural activity from the retina and moved on to neurons throughout the brain. With an estimated 30 per cent of the primate cortex playing some role in visual processing, this was no small undertaking.* In the mid-twentieth century many of the scientists performing these experiments were based in

---

* Primates are admittedly fairly unusual in this sense. Rodent brains, for example, lean more towards processing smell.

the Boston area (many at MIT itself or just north of it, at Harvard) and they were quickly amassing a lot of data that they needed to somehow make sense of.

Perhaps it was the physical proximity. Perhaps it was a tacit acknowledgement of the immense challenge they had each set for themselves. Perhaps in the early days the communities were just too small to keep to themselves. Whatever the reason, neuroscientists and computer scientists forged a long history of collaborating in their attempts to understand the fundamental questions of vision. The study of vision – of how patterns can be found in points of light – is full of direct influence from the biological to the artificial and vice versa. The harmony may not have been constant: when computer science embarked on methods that were useful but didn't resemble the brain, the fields diverged. And when neuroscientists dig into the nitty-gritty detail of the cells, chemicals and proteins that carry out biological vision, computer scientists largely turn away. But the impacts of the mutual influence are still undeniable, and plainly visible in the most modern models and technologies.

★ ★ ★

The earliest efforts to automate vision came before modern computers. Though implemented in the form of mechanical gadgetry, some of the ideas that powered these machines prepared the field for the later emergence of computer vision. One such idea was *template matching*.

In the 1920s, Emanuel Goldberg, a Russian chemist and engineer, set out to solve a problem banks and other offices had while searching their file systems for documents. At the time, documents were stored on microfilm – strips of 35mm film that contained tiny images of documents that could be projected to a larger screen for reading. The ordering of the documents on the film had little relation to their contents, so finding a desired document – such as a cancelled cheque from a particular bank customer – involved much unstructured searching. Goldberg turned to a crude form of 'image processing' to automate this process.

Under Goldberg's plan, cashiers entering a new cheque into the filing system would need to mark it with a special symbol that indicated its contents. For example, three black dots in a row meant the customer's name started with 'A', three black dots in a triangle meant it started with 'B' and so on. Now, if a cashier wanted to find the last cheque submitted by a Mr Berkshire, for example, they just needed to find the cheques marked with a triangle. The triangle pattern was thus a template and the goal of Goldberg's machine was to match it.

Physically, these templates took the form of cards with holes punched in them. So, when looking for Mr Berkshire's documents, the cashier would take a card with three holes punched out in the shape of a triangle and place it in between the microfilm strip and a lightbulb. Each document on the strip would then be automatically pulled up to be aligned with the card, causing the light to shine through the holes on the card and then through the film itself. A photocell placed behind the film detected any light that came through and signalled this to the rest of the machine. For most of the documents, some light would get through as the symbols on the film didn't align with the holes on the card. But when the desired document appeared, the light shining through the card would be exactly blocked out by the pattern of black dots on the film. These mini eclipses meant no light would land on the photocell and this signalled to the rest of the machine, and to the cashier, that a match had been found.

Goldberg's approach required that the cashiers knew in advance exactly what symbol they were looking for and had a card to match it. Crude though it was, this style of template matching became the dominant approach for much of the history of artificial vision. When computers appeared on the scene, the form of the templates migrated from the physical to the digital.

In a computer, images are represented as a grid of pixel values (see Figure 14). Each pixel value is a number indicating the intensity of the colour in the tiny square region of the picture it represents.[*] In the digital world, a template is also just a grid of numbers, one that defines the desired pattern. So, the template for three dots in the shape of a triangle may be a grid of mostly zeros except for three precisely placed pixels with value one. The role of the light shining through the template card in Goldberg's machine was replaced in the computer by a mathematical operation: multiplication. If each pixel value in the image is multiplied by the value at the same location in the template, the result can actually tell us if the image is a match.

---

[*] Actually, pixels in colour images are defined by three numbers corresponding to the intensities of the red, green and blue components. For simplicity, we'll speak of pixels as being a single number, despite the fact that this is only true for grayscale images.

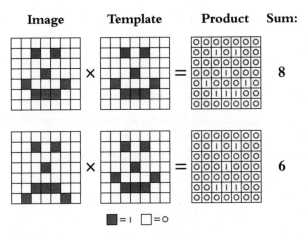

$\blacksquare = 1 \quad \square = 0$

Figure 14

Let's say we are looking for a smiling face in a black and white image (where black pixels have a value of one and white have a value of zero). Given a template for the face, we can compare it with an image through multiplication. If the image is indeed of the face we are searching for, the values that comprise the template will be very similar to those in the image. Therefore, zeros in the template will be multiplied by zeros in the image and ones in the template will be multiplied by ones in the image. Adding up all the resulting values from this multiplication gives us the number of black pixels that are the same in the template and the image, which in the case of a match would be many. If the image we're given is of a frowning face instead, some of the pixels around the mouth in the image won't match the template. There, zeros in the template will be multiplied by ones in the image and vice versa. Because the product at those pixel locations will be zero, the sum across the whole image won't be so high. In this way, a simple sum of products gives a measure of how much an image matches a template.

This method found wide use in many different industries. Templates have been used to count crowd sizes by finding faces in a picture. Known geographical features have also been located in satellite images via templates. The number and model of cars that pass through an intersection can be tracked as well. With template matching, all we need to do is define what we want and multiplication will tell us if it's a match.

★ ★ ★

Imagine a stadium – one just like where you'd watch a football game – but in this stadium, instead of screaming fans, the stands are full of screaming demons. And what they are screaming about isn't players on the field, but rather an image. Specifically, each of these demons has its own preferred letter of the alphabet and when it sees something that looks like that letter on the field it shrieks. The louder the shriek the more the image on the field looks like the demon's favourite letter. Up in the skybox is another demon. This one doesn't look at the field or do any screaming itself, but merely observes all the other demons in the stadium. It finds the demon shrieking the loudest and determines that the image on the field must be that demon's favourite letter.

This is how Oliver Selfridge described the process of template matching at a 1958 conference. Selfridge was a mathematician, computer scientist and associate director of Lincoln Labs at MIT, a research centre focused on national security applications of technology. Selfridge didn't publish many papers himself. He never finished his own PhD dissertation either (he did however end up writing several children's books; presumably these contained fewer demons). Despite this lack of academic output, his ideas infiltrated the research community nonetheless, largely due to the circles in which he moved. After earning a bachelor's degree in mathematics from MIT at only 19 years old, Selfridge was advised in his PhD work by the notable mathematician Norbert Wiener and remained in contact with him. Selfridge also went on to supervise Marvin Minsky, the prominent AI researcher from Chapter 3. And as a graduate student, Selfridge was friends with Warren McCulloch and for a time lived with Walter Pitts (you'll recall this pair of neuroscientists from Chapter 3 as well). Selfridge benefited from letting his ideas marinate in this social stew of prominent scientists.

To map Selfridge's unique analogy to the concept of template matching, we just have to think of each demon as holding its own grid of numbers that represents the shape of its letter. They multiply their grid with the image and sum up those products (just as described above) and scream at a volume determined by that sum. Selfridge doesn't give much of an indication as to why he chose to give such a demonic description of visual processing. His only reflection on it is to say: 'We are not going to apologise for a frequent use of anthropomorphic or biomorphic terminology. They seem to be useful words to describe our notions.'*

---

* Though in response to a colleague's remark on it, Selfridge commented: 'All of us have sinned in Adam, we have eaten of the tree of the knowledge of good and evil, and the demonological allegory is a very old one, indeed.'

Most of the notions in Selfridge's presentation were actually about how the template matching approach is flawed. The demons – each individually checking if their favourite letter was in sight – weren't being very efficient. They each performed their own completely separate computations, but it didn't have to be that way. Many of the shapes that a demon may look for in its search for its letter are also used by other demons. For example, both the 'A'-preferring demon and the 'H'-preferring demon would be on the lookout for a horizontal bar. So why not introduce a separate group of demons, ones whose templates and screams correspond to more basic features of the image such as horizontal bars, vertical lines, slanted lines, dots, etc. The letter demons would then just listen to those demons rather than look at the images themselves and decide how much to scream based on whether the basic shapes of their letter are being yelled about.

From bottom to top Selfridge defined a new style of stadium that contained three types of demon: the 'computational' (those that look at the image and yell about basic shapes), 'cognitive' (those that listen to the computational demons and yell about letters) and 'decision' (the one that listens to the cognitive demons and decides which letter is present). The name Selfridge gave to the model as a whole – to this stack of shrieking demons – was Pandemonium.*

Nefarious nomenclature aside, Selfridge's intuitions on visual processing proved insightful. While conceptually

---

* From the Greek for 'all the demons', introduced in John Milton's *Paradise Lost*.

simple, template matching is practically challenging. The number of templates needed grows with the number of objects you want to be able to spot. If each image needs to be compared to every filter, that's a lot of calculations. Templates also need to more or less match the image exactly. But because of the myriad different patterns of light that the same object can produce on a retina or camera lens, it's nearly impossible to know how every pixel in an image should look when a given object is present. This makes templates very hard to design for any but the simplest of patterns.

These issues make template matching a challenge both for artificial visual systems and for the brain. The ideas on display in Pandemonium, however, represent a more distributed approach, as the features detected by the computational demons are shared across cognitive demons. The approach is also *hierarchical*. That is, Pandemonium breaks the problem of vision into two stages: first look for the simple things, then for the more complex.

Together, these properties make the system more flexible overall. If Pandemonium was set up to recognise the letters of the first half of the alphabet, for example, it would be in a pretty good position to recognise the rest. This is because the low-level computational demons would already know what kinds of basic shapes letters are made of. The cognitive demon for a new letter would just need to figure out the right way to listen to the demons below it. In this way, the elementary features work like a vocabulary – or a set of building blocks – that can be combined and recombined to detect additional complex patterns. Without this hierarchical structure and sharing of low-level features, a basic template-matching approach would need to produce a new template for each letter from scratch.

The design of Pandemonium does pose some questions. For example, how does each computational demon know what basic shape to scream about? And how do the cognitive demons know to whom they should listen? Selfridge proposes that the system learn the answers to these questions through trial and error. If, for example, adjusting how the 'A'-preferring demon listens to those below it makes it better at detecting 'A's, then keep those changes; otherwise don't and try something new. Or, if adding a computational demon to scream about a new low-level pattern makes the whole system better at letter detection, then that new demon stays; otherwise it's out. This is an arduous process of course and it's not guaranteed to work, but when it does it has the desirable effect of creating a system that is customised – automatically – for the type of objects it needs to detect. The strokes that comprise the symbols of the Japanese alphabet, for example, differ from those in the English alphabet. A system that learns would discover the different basic patterns for each. No prior or special knowledge needed, just let the model take a stab at the task.

Computer scientist Leonard Uhr was impressed enough with the ideas of Selfridge and colleagues that he wanted to share their work more broadly. In 1963, he wrote in the *Psychological Bulletin* to an audience of psychologists about the strides computer scientists were making on the vision front. In his article, '"Pattern recognition" computers as models for form perception', he indicates that the models of the time were 'actually already in the position of suggesting physiological and psychological experiments' and even warns that 'it would be unfortunate if psychologists did not play any part in this theoretical development of their own science'. The article is concrete evidence of the intertwined relationship the two fields have always had. But such explicit public pleas for collaboration weren't always needed. Sometimes personal relationships were enough.

Jerome Lettvin was a neurologist and psychiatrist from Chicago, Illinois. He was also a friend of Selfridge, having shared a house with him and Pitts as a young man. A self-described 'overweight slob', Lettvin wanted to be a poet but appeased his mother's wishes and became a doctor instead. The most rebellion he managed was to occasionally abandon his medical practice to engage in some scientific research.

Inspired by the work of his friend and former cohabitant, in the late 1950s Lettvin set out to search for neurons that responded to low-level features – that is, to the types of things computational demons would scream about. The animal he chose to look at was the frog. Frogs use sight mostly to make quick reflexive responses to prey or to predators and as a result their visual system is relatively simple.

Inside the retina, individual light-detecting photore-ceptors send their information to another population of cells called ganglion cells. Each photoreceptor connects to many ganglion cells and each ganglion cell gets inputs from many photoreceptors. But, crucially, all these inputs come from a certain limited region of space. This makes a ganglion cell responsive only to light that hits the retina in that specific location – and each cell has its own preferred location.

At this point in time, ganglion cells were assumed not to do much computation themselves. They were thought of mostly as a relay – just sending information about photoreceptor activity along to the brain like a mail carrier. Such a picture would fit within a template-matching view of visual processing. If the role of the brain was to compare visual information from the eye to a set of stored templates, it wouldn't want that information distorted in any way by the ganglion cells. But if the ganglion cells were part of a hierarchy – where each level played a small role in the eventual detection of complex objects – they should be specialised to detect useful elementary visual patterns. Rather than relaying information verbatim, then, they should be actively processing and repackaging it.

Lettvin found – by recording the activity of these ganglion cells and showing all kinds of moving objects and patterns to the frog – that the hierarchy hypothesis was true. In fact, in a 1959 paper 'What the frog's eye tells the frog's brain' he and his co-authors describe four different types of ganglion cells that each responded to a different simple pattern. Some responded to swift large movements, others to when light turned to dark and still others to curved objects that jittered about. These different categories of responses proved that the ganglion cells were specifically built to detect different elementary patterns. Not only did these findings align with Selfridge's notions of low-level feature detectors, but they also supported the idea that these features are specific to the type of objects the system needs to detect. For example, the last class of cells responded best when a small dark object moved quickly in fits and starts around a fixed background. After describing these in the paper, Lettvin remarked: 'Could one better describe a system for detecting an accessible bug?'

Selfridge's intuitions were proving to be correct. With Lettvin's finding in frogs, the community started to conceive of the visual system more as a stack of screaming demons and less as a store of template cards.

★ ★ ★

Around the same time as Lettvin's work, two doctors at the John Hopkins University School of Medicine in Baltimore were exploring vision in cats. A cat's visual system is more like ours than a frog's. It is tasked with challenging problems related to tracking prey and navigating the environment and is, as a result, more elaborate. The work of the cat visual system is thus stretched over many brain areas and the one that doctors David Hubel* and Torsten Wiesel focused on was the primary visual cortex. This region at the back of the brain represents one of the earlier stages of visual processing in mammals; it gets its input from another brain area – the thalamus – that gets input from the retina itself.

Previous work had investigated how the neurons in the thalamus and retina of cats behave. These cells tend to respond best to simple dots: either a small area of light surrounded by dark or a small area of dark surrounded

---

* Hubel was actually quite interested in mathematics and physics, and was accepted into a PhD programme in physics at the same time he was accepted to medical school. Truly torn, he waited until the last possible day to make the choice.

by light. And, as in the frog, each neuron has a specific location the dot needs to be in for it to respond.

Hubel and Wiesel had access to equipment for producing dots at different locations in order to explore such retinal responses. So, this is the equipment they used, even as they investigated brain areas well beyond the retina. The method for displaying dots included sliding a small glass or metal plate with different cut-out patterns over a screen in front of the eye. Hubel and Wiesel used this to show slide after slide of dots to their feline subject as they measured the activity of a neuron in its primary visual cortex. But the dots simply didn't do it for this neuron – the cell wouldn't make a peep in response to the slides. Then the experimenters noticed something strange: occasionally the neuron would respond – not to the slides – but to the changing of them. As one plate was slid out and another in, the shadow from the edge of the glass swept across the cat's retina. This created a moving line that reliably excited the neuron in the primary visual cortex. One of the most iconic discoveries in neuroscience had just occurred, almost by accident.

Decades later, reflecting on the serendipity of this discovery, Hubel remarked: 'In a certain early phase of science a degree of sloppiness can be a huge advantage.' But that phase quickly passed. By 1960 he and Wiesel had moved their operation to Boston, to help establish the department of neurobiology at Harvard University and embarked on years of careful investigation into the responses of neurons in the visual system.

Expanding on their happy accident, Hubel and Wiesel dug deep into how this responsiveness to moving lines worked. One of their first findings was that the neurons in the primary visual cortex each have a preferred *orientation* in addition to a preferred location. A neuron won't respond to just any line that shows up in its favourite location. Horizontal-preferring neurons require a horizontal line, vertical-preferring neurons require vertical lines, 30-degree-slant-preferring neurons require 30-degree slanted lines, and so on and so on. To get a sense of what this means, you can hold a pen out horizontally in front of your face and move it up and down. You've just excited a group of neurons in your primary visual cortex. Tilt the pen another way and you'll excite a different group (you've now got at-home, targeted brain stimulation for free!).

With their realisation about orientation, Hubel and Wiesel had discovered the alphabet used by the cat brain to represent images. Flies have bug detectors and cats (and other mammals) have line detectors. However, they didn't stop at just observing these responses, they went further to ask how the neurons in the primary visual cortex could come to have such responses. After all, the cells they get their inputs from – those in the thalamus – respond to dots, not lines. So where did the preference for lines come from?

The solution was to assume that neurons in the cortex get a perfectly selected set of inputs from the thalamus. A line, of course, is nothing more than a set of appropriately arranged dots. Inputs to a neuron in the primary visual cortex therefore must come from a set of thalamus neurons wherein each one represents a dot in a row of dots. That way, the primary visual neuron would fire the most when a line was covering all those dots (see Figure 15). Just like the cognitive demons listening for the shrieks of the demons that look for parts of their letter, neurons in the primary visual cortex listen for the activity of neurons in the thalamus that make up their preferred line.

Hubel and Wiesel noticed another kind of neuron, too: ones that also had preferred orientations, but weren't quite as strict about location. These neurons would respond if a line appeared anywhere in a region that was about four times larger than that of the other neurons they recorded. How could these neurons come to have this response? The answer, again, was to assume they got just the right inputs. In particular, a 'complex' neuron – as Hubel and Wiesel labelled these cells – just needed input from a group of regular (or 'simple') neurons. All these simple cells should have the same preferred orientations but slightly different preferred locations. That way, a complex cell would inherit the orientation preference of its inputs, but have a spatial preference that is larger than any single one of them. This spatial flexibility is important. If we want to know if the letter 'A' is in front of us, a little bit of jitter in the exact location of its lines shouldn't really matter. Complex cells are built to discard jitter.

The discovery of complex cells provided an additional piece of the puzzle as to how points of light become perception. In addition to the feature detection done by simple cells, pooling of inputs across space was added to the list of computations performed by the visual system. For all the work they did dissecting this system, Hubel and Wiesel were awarded the Nobel Prize in 1981. In his Nobel lecture, Hubel put their goals plainly: 'Our idea originally was to emphasise the tendency toward

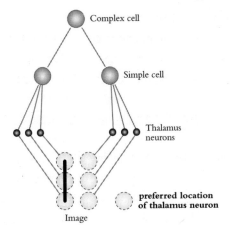

Figure 15

increased complexity as one moves centrally along the visual path, and the possibility of accounting for a cell's behaviour in terms of its inputs.'* This approach, while simple, sufficed to capture many of the basic properties of the visual-processing pathway.

On the other side of the world – at Japan's national public broadcasting organisation, NHK, located in Tokyo – Kunihiko Fukushima heard about the simple properties of the visual system. Fukushima was an engineer and part of the research arm of NHK. Because NHK was a broadcasting company (and was broadcasting visual and audio signals into the eyes and ears of humans)

---

* Hubel and Wiesel did not, however, mention Lettvin or his pioneering work in the frog during this speech. This was an omission Selfridge referred to as 'rotten manners, putting it very mildly'.

it also had groups of neurophysiologists and psychologists on staff to study how sensory signals are received by the brain. These three groups – the psychologists, physiologists and engineers – would meet regularly to share the work of their respective fields. One day, a colleague of Fukushima's decided to present the work of Hubel and Wiesel.

When Fukushima saw this clear description of the roles of neurons in the visual system, he set out to implement the very same functions in a computer model. His model used images of simple white patterns on a black background as input. To approximate the work of the thalamus, a sheet of artificial neurons was created that responded to white dots in the image. This served as a way to get the image information into the network. From here the input to the simple cells needed to be calculated.

To do so, Fukushima used the standard approach of making a grid of numbers that represent the to-be-detected pattern – which in the case of a simple cell is a line with a specific orientation. In engineering terms, this grid of numbers is known as a *filter*. To mimic the spatial preferences of simple cells, Fukushima applied this filter separately at each location in the image. Specifically, the activity of one simple cell was calculated as the sum of the thalamus activity at one location multiplied by the filter. Sliding the filter across the whole image created a set of simple cells all with the same preferred orientation but different preferred locations. This is a process known in mathematics as a *convolution*.

By producing multiple filters – each representing a line with a different orientation – and convolving each with the image, Fukushima produced a full population of simple cells each with its own preferred orientation and location, just like the brain. For the complex cells, he simply gave them strong inputs from a handful of simple cells that all represented the same orientation in nearby locations. That way they would be active if the orientation appeared in any of those locations.

This first version of Fukushima's model was pretty much a direct translation of the physiological findings of Hubel and Wiesel into mathematics and computer code – and, in a way, it worked. It could do some simple visual tasks like finding curved lines in a black and white image, but it was far from a complete visual system and Fukushima knew that. As he later recounted in an interview, after publishing this work in the late 1960s, Fukushima waited patiently to see what Hubel and Wiesel would discover next; he wanted to know what the later stages of visual processing did so he could add them to his model.

But the famous pair of physiologists never provided that information. After their initial work cataloguing cell types, Hubel and Wiesel explored the responses of cells in other visual areas, but were never able to give as clean a description as they had for the primary visual cortex. They eventually moved on to studying how the visual system develops in young animals.

Without the script provided by biology, Fukushima needed to improvise. The solution he devised was to take the structure he had – that of simple cells projecting to complex cells – and repeat it. Stacking more simple and complex cells on top of each over and over creates an extended hierarchy that visual information can be passed through. This means, specifically, a second round of 'simple' cells comes after the initial layer of complex cells. This second layer of simple cells would be on the lookout not for simple features in the image, but rather for simple 'features' in the activity of the complex cells from which they get their input. They'd still use filters and convolutions, but just applied to the activity of the neurons below them. Then these simple cells would send inputs to their own complex cells that respond to the same features in a slightly larger region of space – and then the whole process starts again.

Simple cells look for patterns; complex cells forgive a slight misplacement of those patterns. Simple, complex; simple, complex. Over and over. Repeating this riff leads to cells that are responsive to all kinds of patterns. For a second-layer simple cell to respond to the letter 'L', for example, it just needs to get input from a horizontal-preferring complex cell at one location and a vertical-preferring one at the location just above and to the left of it. A third-layer simple cell could then easily respond to a rectangle by getting input from two appropriately placed 'L'-cells. Go further and further up the chain and cells start responding to larger and more complex patterns, including full shapes, objects and even scenes.

The only problem with extending Hubel and Wiesel's findings this way was that Fukushima didn't actually know how the cells in the different layers should connect to each other. The filters – those grids of numbers that would determine how the simple cells at any given layer respond – had to be filled. But how? For this, Fukushima took a page out of Selfridge's book of Pandemonium and turned to learning.

Rather than using the kind of trial and error Selfridge proposed, Fukushima used a version of learning that doesn't require knowing the right answers. In this form of learning, the model is simply shown a series of images without being told what's in them. The activity of all the artificial neurons is calculated in response to each image and the connections between neurons change depending on how active they are (this may remind you of the Hebbian style of learning discussed in Chapter 4). If a neuron was very active in response to a particular image, for example, the connections from its very active inputs would be strengthened. As a result, that neuron would respond strongly to that and similar images in the future. This makes neurons responsive to specific shapes and different neurons diverge to have different responses. The network is therefore able to pick out a diversity of patterns in the input images.

In the end, Fukushima's model contained three layers of simple and complex cells, and was trained using computer-generated images of the digits zero to four. He dubbed the network the 'Neocognitron' and published the results of it in the journal *Biological Cybernetics* in 1980.

In their original papers, Hubel and Wiesel made a point of stressing that their classification system and nomenclature was not meant to be taken as gospel. The brain is complicated and dividing neurons into only two categories could in no way capture the full diversity of responses and functions. It was just for convenience and expediency of communication that they proceeded in such a way. Yet Fukushima found success in doing the exact thing Hubel and Wiesel warned against: he *did* collapse the brimming complexity of the brain's visual system into two very simple computations. He did take these descriptions as true, or true enough, and even stretched them beyond what they were meant to describe.

This practice – of collapsing and then expanding, of shaking the leaves off a tree and using it to build a house – is what all theorists and engineers know to be necessary if progress is to be made. Fukushima wanted to build a functioning visual system in a computer. Hubel and Wiesel provided a description of the brain's visual system to a first approximation. Sometimes the first approximation is enough.

★ ★ ★

In 1987, like in any other year, the people of Buffalo, New York, mailed countless bills, birthday cards and letters through their local post office. What the citizens of the town didn't know, as they inked the 5-digit zip code of the recipient on to their envelope, was that this bit of their handwriting would be immortalised – digitised and stored on computers across the country for years to come. It would become part of a database for researchers trying to teach computers how to read human handwriting and, in turn, revolutionise artificial vision.

Some of the researchers working on this project were at Bell Labs, a research company owned by the telecommunications company AT&T, located in suburban New Jersey. Among the group of mostly physicists was a 28-year-old French computer scientist named Yann LeCun. LeCun had read about Fukushima and his Neocognitron, and he recognised how the simple repeating architecture of that model could solve many of the hard problems of vision.

LeCun also recognised, however, that the way the model learned its connections needed to change. In particular, he wanted to move back towards the approach of Selfridge and give the model access to images paired with the correct labels of which digit is in them. So, he tweaked some of the mathematical details of the model to make it amenable to a different kind of learning. In this type of learning, if the model misclassifies an image (for example, labels a two as a six), all the connections in the model – those grids of numbers that define what patterns are searched for – are updated in a way that makes them less likely to misclassify that image in the future. In this way, the model learns what patterns are important for identifying digits. This may sound familiar because what LeCun used was the backpropagation algorithm described in Chapter 3. Do this with enough images and the model as a whole becomes quite good at classifying images of handwritten digits, even ones it's never seen before.

LeCun and his fellow researchers unveiled the impressive results of their model, trained on the thousands of Buffalo digits, in 1989. The 'convolutional neural network' – the name given to this style of model – was born.

Just like the template-matching approaches that came before it, convolutional neural networks found applications in the real world. In 1997, these networks formed a core part of a software system AT&T developed to automate the processing of cheques at banks across America. By 2000, it was estimated that between 10 and 20 per cent of cheques in America were being processed by this software. In a charming example of science fulfilling its destiny, Goldberg's dream of equipping banks with synthetic visual systems came true some 70 years after the invention of his microfilm machine.

The method for training convolutional neural networks is a data-hungry one and the model will only learn to be as good as what's fed into it. Just as important as getting the right model, then, is getting the right data. That is why it was so crucial to collect real samples of real digits written by real people. The Bell Lab researchers could've done as Fukushima did and made computer-generated images of numbers. But those would hardly capture the diversity, the nuance or the sloppiness in how digits are written in the wild. The letters that passed through the Buffalo post office contained nearly 10,000 examples of true, human handwriting, giving the model what it needed to truly learn. Once computer scientists saw the importance of real data, they were spurred to collect even more. A dataset of six times as many digits – named MNIST – was collected shortly after the Buffalo set. Surprisingly, this dataset remains one of the most commonly used for quickly testing out new models and algorithms for artificial vision. The digits for MNIST were written by Maryland high school students and US census takers.[*] And while the writers *were* told what their digits were being used for in this case, they almost certainly wouldn't have expected their handwriting to still be used by computer scientists some 30 years later.

Tests of convolutional neural networks didn't stop at digits, but when making the jump to more involved images they hit a snag. In the early 2000s, networks much like LeCun's were trained on another dataset of 60,000 images,

---

[*] You can guess who had the neater handwriting.

this time made of objects. The images were small and grainy – only 32x32 pixels – and could be of either airplanes, cars, birds, cats, deer, dogs, frogs, horses, ships or trucks. While still a simple task for us, this marked a serious increase in difficulty for the networks. The full ambiguity inherent in resolving a three-dimensional world from two-dimensional input comes into play when real images of real objects are used. The same models that could learn to recognise digits struggled to make sense of these more realistic pictures. This brain-like approach to artificial vision was failing at the basic visual processing brains do every day.

A tide turned, however, in 2012 when Alex Krizhevsky, Ilya Sutskever and Geoffrey Hinton from the University of Toronto used a convolutional neural network to win a major image-recognition contest known as the ImageNet Large Scale Visual Recognition Challenge. The contest consisted of labelling images – large (224x224 pixels), real-life pictures taken by people around the world and pulled from image-hosting websites like Flickr – as belonging to one of a thousand different possible object categories. On this very convincing test of visual skill, the convolutional neural network got 62 per cent correct, beating the second-place algorithm by 10 percentage points.

How did the Toronto team do so well? Did they discover some new computation needed for vision? Did they find a magical technique to help the model learn its connections better? The truth, in this case, is actually much more banal. The difference between this convolutional neural network and the ones that came before it was mainly one of size. The Toronto team's network had a total of over 650,000 artificial neurons in it – about 80 times the size of LeCun's digit-recognising one. This network was so large, in fact, that it required some clever engineering to fit the model in the memory of the computer chips that were used to run it. The model went big in another way, too. All those neurons meant a lot more data was needed to train the connections between them. The model learned from 1.2 million labelled images collected by computer science professor Fei-Fei Li as part of the ImageNet database.

A watershed year came in 2012 for convolutional neural networks. While the Toronto team's advances were technically just a quantitative leap – upping the number of neurons and images – the stunning performance enhancement made a qualitative difference in the field. After seeing what they were capable of, researchers flocked to study convolutional neural networks and to make them even better. This usually went in the same direction: making them bigger, but important tweaks to their structure and to how they learned were found as well.

By 2015, a convolutional neural network reached the level of performance expected of a human in the image-classification competition (which isn't actually 100 per cent; some of the images can be confusing). And convolutional neural networks now form the base of almost any image-processing software: facial recognition on social media, pedestrian detection in self-driving cars and even automatic diagnosis of diseases in X-ray images. In an amusing bending of science in on itself, convolutional neural networks have even been used *by neuroscientists* to help automatically detect where neurons are in pictures of brain tissue. Artificial neural networks are now looking at real ones.

It would seem that engineers made a smart move turning to the brain for inspiration on how to build a visual system. Fukushima's attention to the functions of neurons – and his condensing of those functions into simple operations – has paid dividends. But when he was taking the first steps in the development of these models, the computing resources and the data to make them shine simply wasn't available. Decades later, the next generation of engineers picked up the project and brought it across the finish line. As a result, current convolutional neural networks can finally do many of the tasks originally asked for by the MIT summer project in 1966.

But just as Selfridge's Pandemonium helped inspire visual neuroscientists, the relationship between convolutional neural networks and the brain does not go only one way. Neuroscientists have come to reap rewards from the effort computer scientists put into making models that can solve real visual problems. That's because not only are these large, heavily trained convolutional neural networks good at spotting objects in images, they're also good at predicting how the brain will respond to those same images.

★ ★ ★

Visual processing gets started in the primary visual cortex – where Hubel and Wiesel did their recordings – but many areas are involved after that. The primary visual cortex sends connections to (you guessed it) the secondary visual cortex. And after a few more relays, the information ends up in the temporal cortex, located just behind the temples.

The temporal cortex has been associated with object recognition for a long time. As early as the 1930s, researchers noticed that damage to this brain area leads to some strange behaviour. Patients with temporal cortex damage are bad at deciding what things are important to look at and so get easily distracted. They also don't show normal emotional responses to images; they can see pictures most people would find terrifying and hardly blink. And when they want to explore objects, they may do so not by looking at them but by putting them in their mouths.

Understanding of this brain area was refined through decades of careful observation of patients or animals with brain lesions and eventually by recording the activity of its neurons. This led to the conclusion that a subpart of the temporal cortex − the 'inferior' part at the bottom, also called 'IT' − is the main location for object understanding. People with damage to IT have mostly normal behaviour and vision, but with the more specific problem of being unable to appropriately name or recognise objects; they may, for example, fail to recognise the faces of friends or confuse the identity of objects that appear similar.

Accordingly, the neurons in this area respond to objects. Some neurons have clear preferences; one may fire when a clock is present, another for a house, another for a banana, *etc.* But other cells are less scrutable. They may prefer portions of objects or respond similarly to two different objects that have some features in common. Some cells also care about the angle the object is seen from, perhaps firing most if the object is seen straight on, but others are more forgiving and respond to an object at almost any angle. Some care about the size and location of the object, others don't. In total, IT is a grab-bag of neurons interested in objects. While they're not always easy to interpret, such object-driven responses make IT seem like the top of the visual-processing hierarchy, the last stop on the visual system express.

Neuroscientists have tried for decades to understand exactly how IT comes to have these responses. Frequently they followed in the footsteps of Fukushima and built models with stacks of simple and complex cells, hoping that these computations would mimic those that lead to IT activity and make the activity perfectly predictable. This approach worked to an extent, but just like with the Neocognitron, the models were small and they learned their connections from a small set of small images. To make real progress, neuroscientists needed to scale up their models just the same way computer scientists did.

In 2014, two separate groups of scientists – one led by Nikolaus Kriegeskorte at Cambridge University and one by James DiCarlo at MIT – did exactly that. They showed real and diverse images of objects to subjects (humans and monkeys) and recorded the activity of different areas of their visual systems as they viewed them. They also showed the same images to a large convolutional neural network trained to classify real images. What both groups found was that these computer models provided a great approximation to biological vision. Particularly, they showed that if you want to guess how a neuron in IT will respond to a specific image, the best bet – better than any previous method neuroscientists had tried – was to look at how the artificial neurons in the network responded to it. Specifically, the neurons in the last layer of the network best predicted the activity of IT neurons. What's more, the neurons in the second-to-last layer best predicted the activity of neurons in V4 – the area that gives input to IT. The convolutional neural network, it seemed, was mimicking the brain's visual hierarchy.

By showing such a striking alignment between the model and the brain, this research ushered in a revolution in the study of biological vision. It demonstrated that neuroscientists were broadly on the right track, a track that started with Lettvin and Hubel and Wiesel, but that they needed to be bigger and bolder. If they wanted a model that could explain how animals see objects, the model itself needed to be able to see objects.

Going this way, though, symbolised an abandonment of principles that some theorists hold dear: a striving for elegance, simplicity and efficiency in models. There's nothing elegant or efficient about 650,000 artificial neurons wired up in whatever way they found to work. Compared to some of the most beloved and beautiful equations in science, these networks are hulking, unsightly beasts. But, in the end, they work – and there is no guarantee that anything more elegant will.

Selfridge's work pushed biologists to see the visual system as a hierarchy and the experiments that resulted from this planted the seeds for the design of convolutional neural networks. These seeds were incubated in computer science and, in the end, the collaboration yielded fruit for both sides. In general, the desire for artificial systems that can do real visual tasks in the real world has pushed the study of biological vision in directions it may not have gone on its own. Engineers and computer scientists have always enjoyed having the brain's visual system to look to – not only for inspiration, but for proof that this challenging problem is solvable. This mutual appreciation and influence makes the story of the study of vision a uniquely interwoven one.

CHAPTER SEVEN

Cracking the Neural Code

Information theory and efficient coding

Whereas *the heart pumps blood and the lungs effect gas exchange,*
whereas the liver processes and stores chemicals and the kidney
removes substances from the blood, the nervous system processes
information.

Summary of Neurosciences Research
Program work session, 1968

The goal of the 1968 Neurosciences Research Program
meeting was to discuss how individual and groups of
neurons process information. The meeting's summary,
written by neuroscientists Theodore Bullock and Donald
Perkel, does not push for any hard and fast conclusions.
But it does lay out a wide world of possibilities for the
representation, transformation, transmission and storage
of information in the brain in a way that summarised the
state of the field.

As the quote from their summary implies, ascribing
the role of information processing to the brain seems as
natural as saying the heart pumps blood. Even before
'information' became a part of everyday vocabulary in
the twentieth century, scientists still spoke implicitly of
the information that nerves convey, often in the language
of 'messages' and 'signals'. An 1892 lecture to hospital
employees, for example, explains that: 'There are fibres

which convey messages from the various parts of the body to the brain' and that some of these fibres 'carry special kinds of messages as, for example, the nerves connected with the organs of special sense, which have been called the gateways of knowledge'. In the same vein, an 1870 publication describes the firing of motor neurons as 'a message of the will to the muscle' and even goes so far as to equate the nervous system with the dominant information-transmitting technology of the day: the telegraph.

But the investigation into how the nervous system represents information only started in earnest about 40 years before Bullock and Perkel's report, with the work of Edgar Adrian in the early twentieth century.

Adrian was in many ways the image of a prim and proper man of science. By the time he was born in London in 1889, his family had been in England for more than 300 years – a lineage that included a sixteenth-century surgeon and several reverends and members of government. As a student, his brilliance was regularly acknowledged by his teachers. In addition to his focus on medicine during his university studies, he displayed skill in art, particularly painting and drawing. As a lecturer at Cambridge, he worked long hours in the lab and in the classroom. In his career as a physiologist, he was an undeniable success. At the age of 42 he won a Nobel Prize and in 1955 he was granted a title by Queen Elizabeth II, becoming Lord Adrian.

But behind these formal awards and accolades was a restless and chaotic man. Adrian was a thrill-seeker who liked climbing mountains and driving fast cars. He was happy to experiment on himself, including keeping a needle in his arm for two hours to try to measure muscle activity. He was known to play elaborate games of hide-and-seek with fellow students in the valleys of England's Lake District. As a professor he was equally elusive. He avoided unscheduled meetings by hiding in his lab, forcing any enquiring students to try to catch him on his bike ride home. He was temperamental and when he needed to think he'd perch himself on a shelf in a dark cabinet. His lab mates and his family described his movements as rapid, jerky and almost constant. His mind could be equally darting. Over the course of his career he studied many different questions in many different animals: vision, pain, touch and muscle control in frogs, cats, monkeys and more.

This inability to remain still, physically or mentally, may have been a key to his success. Through his varied studies on the activity of single nerves he was able to find certain general principles that would form the core of our understanding of the nervous system as a whole. In his 1928 book, *The Basis of Sensation*, Adrian explains his conclusions and the experiments that allowed him to reach them. The pages are peppered with talk of 'signals', 'messages' and even 'information', all mixed in with anatomical details of the nervous system and the technical challenges of capturing its activity. It was a mix of experimental advances and conceptual insights that would influence the field for decades to come.

In Chapter 3, Adrian explains an experiment wherein he adds weight to a frog's muscle to see how the 'stretch' receptors that track the muscle's position would respond. Adrian recorded from the nerves that carry this signal from the receptors to the spinal cord. After applying different weights, Adrian summarised his findings as follows: 'The sensory message which travels to the central nervous system when a muscle is stretched ... consists of a succession of impulses of the familiar type. The frequency with which the impulses recur depends on the strength of the stimulus, but the size of each impulse does not vary.' This finding – that the size, shape or duration of an action potential emitted by these sensory neurons does not change, no matter how heavy or light the weight applied to the muscle is – Adrian referred to as the 'all-or-nothing' principle.

Examples of the 'all-or-nothing' nature of neural impulses reappear throughout the book. In different species, for different nerves carrying different messages, the story is always the same. Action potentials don't change based on the signal they're conveying, but their frequency can. The spikes of a neuron are thus like an army of ants – each as identical as possible, their power coming mainly from their numbers.

If the nature of an individual action potential is the same regardless of the strength or weakness of the sensory stimulus causing it, then one thing is certain: the size of the action potential does not carry information. With the contributions from Adrian, physiologists now felt comfortable embarking on a search for where exactly information was in nerves and how it got transmitted.

There was only one problem: what *is* information? The blood that the heart pumps and the gases that lungs exchange are real, physical substances. They're observable, tangible and measurable. For as commonly as we use it, 'information' is actually a rather vague and elusive concept. A precise definition of the word does not easily come to mind for most people; it falls unfortunately into the 'know it when you see it' camp. Without a way to weigh information the way we can weigh fluids or gases, what hope could scientists have for a quantitative understanding of the brain's central purpose?

Between the time of Adrian's book and Perkel and Bullock's report, however, a quantitative definition of information had been found. It was born out of the scientific struggles of the Second World War and went on to transform the world in unexpected ways. Its application to the study of the brain was at times as rocky to execute as it was obvious to attempt.

★ ★ ★

Claude Shannon started at Bell Labs under a contract provided by the American military. It was 1941 and the National Defense Research Committee wanted scientists working on wartime technology. The seriousness of the work didn't dampen Shannon's naturally playful tendencies, though. He enjoyed juggling and while at Bell Labs was known to juggle around campus while riding a unicycle.

Born in a small town in the American Midwest, Shannon grew up following his curiosity about all things science, mathematics and engineering anywhere it took him. As a child he played with radio parts and enjoyed number puzzles. As an adult he created a mathematical theory of juggling and a flame-powered Frisbee. He enjoyed chess and building machines that could play chess. A constant tinkerer, he made many gadgets, some more productive than others. On his desk at Bell Labs, for example, he kept an 'Ultimate Machine': a box with a switch that, when flipped on, caused a mechanical hand to reach out and flip it back off.[*]

For his master's degree, Shannon wrote a 72-page thesis entitled 'A symbolic analysis of relay and switching circuits' that would revolutionise electrical engineering. For his PhD, he turned his mathematical eye toward biology, working on 'An Algebra for Theoretical Genetics'. But his topic at Bell Labs was cryptography. How to safely encode messages that would be transmitted through land, air and water was a natural topic of concern for the military. Bell Labs was a hub of cryptography research and even hosted the renowned code-cracker Alan Turing during Shannon's time there.

All this work on codes and messages got Shannon thinking broadly about the concept of communication. During the war, he proposed a method for understanding message-sending mathematically. Because of the necessary secrecy around cryptography research, however, his ideas were kept classified. In 1948, Shannon was finally able to publish the work and 'A mathematical theory of communication' became the founding document of a new field: information theory.

Shannon's paper describes a very generic communication system consisting of five simple parts. The first is an information *source*, which produces the message that will be sent. The next is the *transmitter*,

---

[*] Marvin Minsky, one of the authors of the *Perceptrons* book from Chapter 3, was working under Shannon at the time and is credited with the design of the Ultimate Machine. Shannon reportedly convinced Bell Labs to produce several of them as gifts for AT&T executives.

Shannon's Communication System

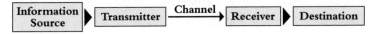

Figure 16

which is responsible for encoding the message into a form that can be sent across the third component, the *channel*. On the other end of the channel, a *receiver* decodes the information back into its original form and it is sent to its final *destination* (Figure 16).

In this framework, the medium of the message is irrelevant. It could be songs over radio waves, words on a telegraph or images through the internet. As Shannon says, the components of his information-sending model are 'suitably idealised from their physical counterparts'. This is possible because, in all of these cases, the fundamental problem of communication remains the same. It is the problem of 'reproducing at one point either exactly or approximately a message selected at another point'.

With this simple communication system in mind, Shannon aimed to formalise the study of information transmission. But, to mathematically approach the question of how information is communicated, he first had to define information mathematically. Building on previous work, Shannon discusses what desirable properties a measure of information should have. Some are practical: information shouldn't be negative, for example, and its definition should be easy to work with mathematically. But the real constraint came from the need to capture an intuition about information: its reliance on a code.

Imagine a school where all students wear uniforms. Seeing a student show up in the same outfit every day provides very little information about their mood, their personality or the weather. On the other hand, in a school without uniforms, clothing choice has the ability to convey all this information and more. To someone wondering about the current temperature, for example, seeing a student in a sundress instead of a sweater can go a long way towards relieving that curiosity. In this way, clothing can be used as a code – it is a transmittable set of symbols that conveys meaning.

The reason the uniformed students can't carry this information is that a code requires options. There need to be multiple symbols in a code's vocabulary (in this case, multiple outfits in a student's wardrobe), each with their own meaning, for any of the symbols to have meaning.

But it's not just the number of symbols in a code that matters – it's also how they're used. Let's say a student has two outfits: jeans and a T-shirt or a suit. If the student wears jeans and a T-shirt 99 per cent of the time, then there is not much information that can be gained from this wardrobe choice. You wouldn't even need to see the student to be almost certain of what they're wearing – it's essentially a uniform. But the one day in a hundred where they show up in a suit tells you something important. It lets you know that day is somehow special. What this shows is that the rarer a symbol's use, the more information it contains. Common symbols, on the other hand, can't communicate much.

Shannon wanted to capture this relationship between a symbol's use and its information content. He therefore defined a symbol's information content in terms of the probability of it appearing. Specifically – to make the amount of information decrease as the probability of the symbol increases – he made a symbol's information depend on the *inverse* of its probability. Because the inverse of a number is simply one divided by that number, a higher probability means a lower 'inverse probability'. In this way, the more frequently the symbol is used, the lower its information will be. Finally, to meet his other mathematical constraints, he took the logarithm of this value.

A logarithm, or 'log', is defined by its *base*. To take the base-10 log of a number, for example, you would ask: 'To what power must I raise 10 in order to get this number?' The base-10 log of 100 (written as $\log_{10}100$), is therefore 2, because 10 to the power of 2 (*i.e.*, 10x10) is 100. The base-10 log of 1,000 is thus 3. And the base-10 log of something in between 100 and 1,000 is in between 2 and 3.

Shannon decided to use a base of *two* for his definition of information. To calculate the information in a symbol you must therefore ask: 'To what power must I raise two in order to get the inverse of the symbol's probability?' Taking our student's outfit of jeans and T-shirt as a symbol that appears with 0.99 probability, its information content is $\log_2(1/0.99)$, which comes out to about 0.014. The suit that appears with only 0.01 probability, on the other hand, has an information content of $\log_2(1/0.01)$ or roughly 6.64. Again, the lower the probability, the higher the information.[*]

---

[*] More on probability and its history in Chapter 10.

But Shannon was interested in more than just the information in a single symbol – he wanted to study the information content of a code. A code is defined by its set of symbols and how frequently each is used. Shannon therefore defined the total information in a code as the sum of the information of all its symbols. Importantly, this sum is weighted – meaning the information from each symbol is multiplied by how frequently that symbol is used.

Under this definition, the student's clothing code would have a total information content of 0.99 x 0.014 (from the jeans and T-shirt) + 0.01 x 6.64 (from the suit) = 0.081. This can be thought of as the average amount of information we would receive each day by seeing the student's outfit. If the student chose instead to wear their jeans 80 per cent of the time and their suit the other 20 per cent, their code would be different. And the average information content would be higher: $0.80 \times \log_2(1/0.80) + 0.20 \times \log_2(1/0.20) = 0.72$.

Shannon gave the average information rate of a code a name. He called it entropy. The official reason he gives for this is that his definition of information is related to the concept of entropy in physics, where it serves as a measure of disorder. On the other hand, Shannon was also known to claim – perhaps jokingly – that he was advised to call his new measure entropy because 'no one understands entropy' and therefore Shannon would likely always win arguments about his theory.

Shannon's entropy captures a fundamental trade-off inherent in maximising information. Rare things carry the most information, so you want as much of them as possible in your code. But the more you use a rare symbol, the less rare it becomes. This fight fully defines the equation for entropy: decreasing the probability of the symbol makes the log of its inverse go up – a positive contribution to the information. But this number is then multiplied by that very same probability: this means that decreasing a symbol's probability makes its contribution to information go down. To maximise entropy, then, we must make rare symbols as common as possible but no commoner.

Shannon's use of a base-two log makes the unit of information the *bit*. Bit is short for binary digit and, while Shannon's paper sees the first known use of the word, he did not coin it (Shannon credits his Bell Labs colleague John Tukey with that honour).* The bit as a unit of information has a helpful and intuitive interpretation. Specifically, the average number of bits in a symbol is equal to the number of yes-or-no questions you need to ask in order to get that amount of information.

For example, consider trying to find out the season in which someone was born. You may start by asking: 'Is it a transitional season?' If they say yes, you may then ask: 'Is it spring?' If they say yes to that, you have your answer; if they say no, you still have your answer: autumn. If they said no to the first question you'd follow the opposite route – asking if they were born in summer, *etc.* No matter the answer, it takes two yes-or-no questions to get it. Shannon's entropy equation agrees. Assuming

---

* The base-two log wasn't the only option for information. Prior to Shannon's work, his colleague Ralph Hartley posited a definition of information using a base-10 log, which would've put information in terms of 'decimal digits or 'dits' instead of bits.

people are equally likely to be born in each season, then each of these season 'symbols' will be used 25 per cent of the time. The information in each symbol is thus $\log_2(1/0.25)$. This makes the average bits per symbol equal to two – the same as the number of questions.

Part of designing a good communication system is designing a code that packs in a lot of information per symbol. To maximise the average information that a symbol in a code provides, we need to maximise the code's entropy. But, as we saw, the definition of entropy has an inherent tension. To maximise it, rare symbols need to be the norm. What is the best way to satisfy this seemingly paradoxical demand? This tricky question turns out to have a simple answer. To maximise a code's entropy, each of its symbols should be used the exact same amount. Have five symbols? Use each one-fifth of the time. A hundred symbols? Each should have a probability of 1/100th. Making each and every symbol equally likely balances the trade-off between rare and common communication.

What's more, the more symbols a code has the better. A code with two symbols, each used half the time, has an entropy of one bit per symbol (this makes sense according to our intuitive definition of a bit: thinking of one symbol as standing in for 'yes' and the other for 'no', each symbol answers one yes-or-no question). On the other hand, a code with 64 symbols – each used equally – has an entropy of six bits per symbol.

As important as a good code is, encoding is only the start of a message's journey. According to Shannon's conception of communication, after information is encoded it still needs to be sent through a channel on the way to its destination. Here is where the abstract aims of message-sending meet the physical limits of matter and materials.

Consider the telegraph. A telegraph sends messages via short pulses of electric current passed over wires. The patterns of pulses – combinations of shorter 'dots' and longer 'dashes' – define the alphabet. In American Morse code, for example, a dot followed by a dash indicates the letter 'A', two dots and a dash means 'U'. The physical constraints and imperfections of the wires carrying these messages, especially those sent long distances or under oceans, put a limit on the speed of information. Telegraph operators who typed too quickly were at risk of running their dots and dashes together, creating an unintelligible 'hog Morse' that would be worthless to the receiver of it. In practice, operators could safely send around 100 letters per minute on average.

To create a practical measure of information rate, Shannon combined the inherent information rate of a code with the physical transmission rate of the channel it is sent over. For example, a code that provides five bits of information per symbol and is sent over a channel that can send 10 symbols per minute would have a total information rate of 50 bits per minute. The maximum rate at which information can be sent over a channel without errors is known as the channel's capacity.

Shannon's publication enforced a clear structure on a notoriously nebulous concept. In this way, it set the stage for the increasing objectification of information in the decades to come. The immediate effects of Shannon's work on information processing in the real world, however, were slight. It took more than two decades for the technology that makes information transmission, storage and processing a constant part of everyday life to come about. And it took engineers time to figure out how to harness Shannon's theory for practical effect in these devices. Information theory's impact on biology, however, came much quicker.

★ ★ ★

The first application of information theory to biology was itself a product of war. Henry Quastler, an Austrian physician, was living and working in the US during the time of the Second World War. His response to the development of the atomic bomb was one of horror – and action. He left his private practice to start doing research on the medical and genetic effects of nuclear bombs. But he needed a way to quantify just how the information encoded in an organism was changed by exposure to radiation. 'A godsend, these formulas, splendid! I can go on now,' Quastler is said to have remarked upon learning about Shannon's theory. He wrote a paper in 1949 – just a year after Shannon's work was published – entitled 'The information content and error rate of living things'. It set off the study of information in biology.

Neuroscience was not slow to follow. In 1952, Warren McCulloch and physicist Donald MacKay published 'The limiting information capacity of a neuronal link'. In this paper, they derive what they consider to be the most optimistic estimate of how much information a single neuron could carry. Based on the average time it takes to fire an action potential, the minimum time needed in between firing and other physiological factors, MacKay and McCulloch calculated an upper bound of 2,900 bits per second.

MacKay and McCulloch were quick to stress that this doesn't mean neurons actually *are* conveying that much information, only that under the best possible circumstances they could. After their paper, many more publications followed, each aiming to work out the true coding capacity of the brain. So awash in attempts was the field that in 1967 neuroscientist Richard Stein wrote a paper both acknowledging the appeal of information theory for quantifying nervous transmission but also lamenting the 'enormous discrepancies' that have resulted from its application. Indeed, in the work that followed MacKay and McCulloch's, estimates ranged from higher than their value – 4,000 bits per second per neuron – to significantly lower, as meagre as one-third of a bit per second.

This diversity came, in part, from different beliefs about how the parts and patterns of nervous activity should be mapped on to the formal components of Shannon's information theory. The biggest question centred on how to define a symbol. Which aspects of neural activity are actually carrying information and which are incidental? What, essentially, is the neural code?

Adrian's original finding – that it is not the height of the spike that matters – still held strong.* But even under this constraint, options abounded. Starting from the basic

---

* Full disclosure: some modern neuroscientists are exploring the idea that action potentials actually *do* change in certain ways depending on what inputs the cell gets and that these changes *could* be part of the neural code. Science is never set in stone.

unit of an action potential, scientists were still able to devise many conceivable codes. MacKay and McCulloch began by thinking of the neural code as composed of only two symbols: spike or no spike. At each point in time a neuron would send one or the other symbol. But after calculating the information rate of such a code, MacKay and McCulloch realised they could do better. Thinking instead of the *time between spikes* as the code allowed a neuron to transmit much more information. In this coding scheme, if there were a 20-millisecond gap in between two spikes, this would symbolise something different from a 10-millisecond gap. It's a scheme that creates many more possible symbols and it was with this style of coding that they made their estimate of 2,900 bits per second.

Stein, in his attempt to clean up the cacophony of codes on offer at the time, focused on a third option for neural coding – the one that came from Adrian himself. Adrian, after establishing that action potentials don't change as the stimulus does, claimed that: 'In fact, the only way in which the message can be made to vary at all is by a variation in the total number of the impulses and in the frequency with which they recur.' This style of coding – where it is the number of spikes produced in a certain amount of time that serves as the symbol – is known as frequency or rate-based coding. In his 1967 paper, Stein argues for the existence of a rate-based code and highlights its benefits, including higher error tolerance.

But the debate over what the true neural code is did not end with Stein in 1967. Nor did it end with Bullock and Perkel's meeting on information coding in the brain a year later. In fact, in their report on that meeting, Bullock and Perkel include an Appendix that lays out dozens of possible neural codes and how they could be implemented.

In truth, neuroscientists continue to spar and struggle over the neural code to this day. They host conferences centred on 'Cracking the neural code'. They write papers with titles like 'Seeking the neural code', 'Time for a new neural code?' and even 'Is there a neural code?' They continue to find good evidence for Adrian's original rate-based coding, but also some against it. Identifying the neural code can seem a more distant goal now than when MacKay and McCulloch wrote their first musings on it.

In general, some evidence of rate-based coding can be found in most areas of the brain. The neurons that send information from the eye change their firing frequency based on the intensity of light. Neurons that encode smell fire in proportion to the concentration of their preferred odour. And, as Adrian showed, receptors in the muscle and those in the skin fire more with more pressure applied to them. But some of the strongest evidence for other coding schemes comes from sensory problems that require very specific solutions.

When localising the source of a sound, for example, precise timing matters. Because of the distance between the two ears, sound coming from the left or right side will hit one ear just before it hits the other. This gap between the times of arrival at each ear – sometimes as brief as only a few millionths of a second – provides a clue for calculating where the sound came from. The medial superior olive (MSO), a tiny cluster of cells located right in between the two ears, is responsible for performing this calculation.

The neural circuit that can carry this out was posited by psychologist Lloyd Jeffress in 1948 and has been supported by many experiments since. Jeffress' model starts with information coming from each ear in the form of a temporal code – that is, the exact timing of the spikes matters. In the MSO, cells that receive inputs from each ear compare the relative timing of these two inputs. For example, one cell may be set up to detect sounds that arrive at both ears simultaneously. To do so, the signals from each ear would need to take the exact same amount of time to reach this MSO cell. This cell then fires when it receives two inputs at the exact same time and this response indicates that the sound hit both ears at the same time (see Figure 17).

The cell next to this one, however, receives slightly asymmetric inputs. That is, the nerve fibre from one ear needs to travel a *little* farther to reach this cell than the nerve from the other ear. Because of this, one of the temporal signals gets delayed. The extra length the signal travels determines just how much extra time it takes. Let's say the signal from the left ear takes an extra 100 microseconds to reach this MSO cell. Then, the only way this cell will receive two inputs at once is if the sound hits the left ear 100 microseconds before it hits the right. Therefore, this cell's response (which, like the other cell, only comes when it receives two inputs at once) would signal a 100-microsecond difference.

Continuing this pattern, the next cell may respond to a 200-microsecond difference, the one after that 300 microseconds and so on. In total, the cells in the MSO form a map wherein those that signal short arrival time

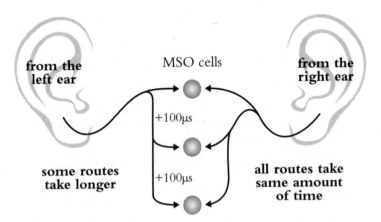

Figure 17

differences fall at one end and those that signal long differences fall at the other. In this way, a temporal code has been transformed into a *spatial* code: the position of the active neuron in this map carries information about the source of the sound.

For the question of why the neural code is such an enigma, the most likely answer – as with so many questions of the brain – is because it's complicated. Some neurons, in some areas of the brain, under some circumstances, may be using a rate-based code. Other neurons, in other times and places, may be using a code based on the timing of spikes, or the time in between spikes, or some other code altogether. As a result, the thirst to crack *the* neural code will likely never be quenched. The brain, it seems, speaks in too many different languages.

★ ★ ★

Evolution did not furnish the nervous system with one single neural code, nor did it make it easy for scientists to find the multitude of symbols it uses. But, according to British neuroscientist Horace Barlow, evolution did thankfully provide one strong guiding light for our understanding of the brain's coding scheme. Barlow is known as one of the founders of the efficient coding hypothesis, the idea that – no matter what code the brain uses – it is always encoding information *efficiently*.

Barlow was a trainee of Lord Adrian. He worked with him – when he could find him – as a student at Cambridge in 1947. Barlow had always had a keen interest in physics and mathematics but, to be practical, chose to study medicine.* Yet throughout his studies he recognised how the influence from more quantitative subjects could drive questions in biology. It was a trait he considered in contrast to his mentor: '[Adrian] was not at all theoretically based; his attitude was that we had the means of recording from nerve fibres and we should just see what happens.'

Quickly taken in by Shannon's equations when they came about, Barlow made several early contributions to the study of information in the brain. Rather than simply counting bits per second, however, Barlow's use of information theory went deeper. The laws of information, in some respects, are as fundamental and constraining of biology as the laws of physics. From Barlow's perspective, these equations could thus do more than merely *describe* the brain as it is, but rather *explain* how it came to be. So certain of its importance to neuroscience, Barlow compared trying to study the brain without focusing on

---

* Barlow credits his mother Nora, granddaughter of Charles Darwin, with his interest in science.

information processing to trying to understand a wing without knowing that birds fly.

Barlow came to his efficient coding hypothesis by combining reflections on information theory with observations of biology. If the brain evolved within the constraints of information theory – and evolution tends to find pretty good solutions – then it makes sense to conclude that the brain is quite good at encoding information. 'The safe course here is to assume that the nervous system is efficient,' Barlow wrote in a 1961 paper. If this is true, any puzzle about why neurons are responding the way they are may be solved by assuming they are acting efficiently.

But what does efficient information coding look like? For that, Barlow focused on the notion of redundancy. In Shannon's framework, 'redundancy' refers to the size of the gap between the highest possible entropy a given set of symbols could have and the entropy they actually have. For example, if a code has two symbols and uses one of them 90 per cent of the time and the other only 10 per cent, its entropy is not as high as it could be. Sending the same symbol nine out of ten times is redundant. As we saw earlier, the code with the highest entropy would use each of those symbols 50 per cent of the time and would have a redundancy of zero. Barlow believed efficient brains reduce their redundancy as much as possible.

The reason for this is that redundancy is a waste of resources. The English language, as it turns out, is incredibly redundant. A prime example of this is the letter 'q', which is almost always followed by 'u'. The 'u' adds little if any information once we see the 'q' and is therefore redundant. The redundancy of English means we could, in theory, be conveying the same amount of information with far fewer letters. In fact, in his original 1948 paper, Shannon estimated the redundancy of written English to be about 50 per cent. This is why, for example, ppl cn stll rd sntncs tht hv ll th vwls rmvd.*

In the nervous system, redundancy can come in the form of multiple neurons saying the same thing. Imagine one neuron represents the letter 'q' and another the letter 'u'. The sight of 'qu' would thus make both these neurons fire. But if these two letters frequently appear together in the world, it would be more efficient of the brain to use just a single neuron to respond to them.

Why should it matter if the brain is efficient with its encoding? One reason is energy costs. Every time a neuron fires a spike, the balance of charged particles inside and outside the cell gets thrown off. Restoring this balance takes energy: little pumps in the cell membrane have to chuck sodium ions out of the cell and pull potassium ones back in. Building up neuro-transmitters and expelling them from the cell with each spike also incurs a cost. In total, it's estimated that up to three-fourths of the brain's energy budget goes towards sending and receiving signals. And the brain – using 20 per cent of the body's energy while accounting for only 2 per cent of its weight – is the most energetically expensive organ to run. With such a high energy bill, it

---

* In case you couldn't: 'people can still read sentences that have all the vowels removed'. Texting and tweeting are also great ways to see just how many letters can be removed from a word before causing problems.

makes sense for the brain to be economical in how it uses its spikes.

But to know how to send information efficiently, the brain needs to know what kind of information it normally needs to send. In particular, the brain needs to somehow determine when the information it is receiving from the world is redundant. Then it could simply not bother sending it on. This would keep the neural code efficient. Does the nervous system have the ability to track the statistics of the information it's receiving and match its coding scheme to the world around it? One of Lord Adrian's own findings – adaptation – suggests it does.

In his experiments on muscle stretch receptors, Adrian noticed that 'there is a gradual decline in the frequency of the discharge under a constant stimulus'. Specifically, while keeping the weight applied to the muscle constant, the firing rate of the nerve would decrease by about half over 10 seconds. Adrian called this phenomenon 'adaptation' and defined it as 'a decline in excitability caused by the stimulus'. Noticing the effect in several of his experiments, he devoted a whole chapter to the topic in his 1928 book.

Adaptation has since been found all over the nervous system. For example, the 'waterfall effect' is a visual illusion wherein the sight of movement in one direction then causes stationary objects to appear as though they are moving in the opposite direction. It's so-named because it can happen after staring at the downward motion of a waterfall. The effect is believed to be the result of adaptation in the cells that represent the original motion direction: with these cells silenced by adaptation, our perception is biased by the firing of cells that represent the opposite direction.

In his 1972 paper, Barlow argues for adaptation as a means of increasing efficiency: 'If sensory messages are to be given a prominence proportional to their informational value, mechanisms must exist for reducing the magnitude of representation of patterns which are constantly present, and this is presumably the underlying rationale for adaptive effects.'

In other words – specifically, in the words of information theory – if the same symbol is being sent across the channel over and over, its presence no longer carries information. Therefore, it makes sense to stop sending it. And that is what neurons do: they stop sending spikes when they see the same stimulus over and over.

Since the time that Barlow made the claim that cells should adapt their responses to the signals they're receiving, techniques for tracking how neurons encode information have developed that allow for more direct and nuanced tests of this hypothesis. In 2001, for example, computational neuroscientist Adrienne Fairhall, along with colleagues at the NEC Research Institute in Princeton, New Jersey, investigated the adaptive abilities of visual neurons in flies.

For their experiment, the researchers showed the flies a bar moving left and right on a screen. At first, the bar's motion was erratic. At one moment it could be moving very quickly leftwards, and at the next it could go equally fast towards the right, or it could stay in that direction, or it could slow down entirely. In total, its range of possible speeds was large. After several seconds of such chaos, the bar then calmed down. Its movement became more constrained, never going too quickly in either direction.

Over the course of the experiment, the bar switched between such periods of erratic and calm movement several times.

Looking at the activity of the neurons that respond to motion, the researchers found that the visual system rapidly adapts its code to the motion information it's currently getting. Specifically, to be an efficient encoder, a neuron should always fire at its peak firing rate for the fastest motion it sees and at its lowest for the slowest.[*] Thinking of different rates of firing as different symbols in the neural code, spreading the firing rates out this way ensures that all these symbols get used roughly equally. This maximises the entropy of the code.

The problem is that the fastest motion during the period of calm is much slower than the fastest motion during the more erratic period. This means that the same exact speed needs to map to two different firing rates depending on the context it appears in. Strange as it is, this is just what Fairhall and colleagues saw. During the calm period, when the bar was moving at its fastest speed, the neuron fired at over 100 spikes per second. Yet when that same speed occurred during the erratic period, the neuron only fired about 60 times per second. To get the neuron back up to 100 spikes per second during the erratic period, the bar needed to move 10 times as fast.

In addition, the researchers were able to quantify the amount of information carried by a spike before and after

---

[*] Technically if the neuron has a preferred direction of motion – that is, it fires most strongly for, say, rightwards motion – it should fire at its peak rate for high speed in that direction and its lowest rate for high speed in the opposite direction. But the principle remains the same regardless.

the switch between these two types of motion. During the erratic period, the information rate was around 1.5 bits per spike. Immediately after the switch to calm movement, it dropped to just 0.8 bits per spike: the neuron, having not yet adapted to the new set of motion it was seeing, was an inefficient encoder. After just a fraction of a second of exposure to the calmer motion, however, the bits per spike went right back up to 1.5. The neuron needed just a small amount of time to monitor the range of speeds it was seeing and adapt its firing patterns accordingly. This experiment shows that, just as Barlow's efficient coding theory suggests, adaptation ensures that all types of information are encoded efficiently.

Neuroscientists also believe that the brain is built to produce efficient encodings on much longer timescales than the seconds to minutes of a sensory experience. Through both evolution and development, an organism has a chance to sample its environment and adapt its neural code to what is most important to it. By assuming that a given area of the brain is best-suited to represent relevant information as efficiently as possible, scientists are attempting to reverse engineer the evolutionary process.

The 30,000 nerves that leave the human ear, for example, respond to different types of sounds. Some of the neurons prefer short blips of high-pitched noises, others prefer low-pitched noises. Some respond best when a soft sound gets louder, others when a loud sound gets softer and others still when a soft sound gets louder then softer again. Overall, each nerve fibre has a complex pattern of pitches and volumes that best drive its firing.

Scientists know, for the most part, *how* the fibres end up with these responses. Tiny hairs connected to cells in the inner ear move in response to sounds. Each cell responds to a different pitch based on where it is in a small, spiral-shaped membrane. The nerve fibres that leave the ear get inputs from these hairy cells. Each fibre combines pitches in its own way to make its unique, combined response profile.

What is less clear, however, is *why* the fibres have these responses. That's where ideas from information theory can help.

If the brain is indeed reducing redundancy as Barlow suggests, then only a small number of neurons should be active at a time. Neuroscientists refer to this kind of activity as 'sparse'.* In 2002, computational neuroscientist Michael Lewicki asked whether the response properties of auditory nerves could be the result of the brain enforcing a sparse code – one specifically designed for the sounds an animal needs to process.

To answer this, he first had to gather a collection of different natural sounds. One set of sounds came from a CD of vocalisations made by rainforest animals such as bats, manatees and marmosets; another was a compilation of 'background' noises like crunching leaves and snapping twigs; and the third was from a database of human voices reading English sentences.

Lewicki then used an algorithm to decompose these complex sounds into a dictionary of short sound patterns.

---

* Among neuroscientists, the 'grandmother cell' is considered the mascot of sparse coding. This fictional neuron is meant to be the one and only cell to fire when you see your grandmother (and fires in response to nothing else). Such an extreme example of efficient coding was devised by Jerome Lettvin (the frog guy from the last chapter) in order to vividly demonstrate the concept to his students.

The goal of the algorithm was to find the best decomposition – that is, one that can recreate each full, natural sound using as few of the short sound patterns as possible. In this way, the algorithm was seeking a sparse code. If the brain's auditory system evolved to sparsely encode natural sounds, the sound patterns preferred by the auditory nerves should match those found by the algorithm.

Lewicki found that creating a dictionary from just the animal noises alone produced sound patterns that didn't match the biology. Specifically, the patterns the algorithm produced were too simple – representing just pure tones rather than the complex mix of pitches and volumes that human and animal auditory nerves tend to prefer. Applying the algorithm to a mix of animal noises and background sounds, however, did mimic the biology. This suggests that the coding scheme of the auditory system is indeed matched to these environmental sounds, allowing it to encode them efficiently. What's more, Lewicki found that a dictionary made from human speech also reproduced the sound profiles preferred by biology. Lewicki took this as evidence for the theory that human speech evolved to make best use of the existing encoding scheme of the auditory system.[*]

★ ★ ★

In 1959, Barlow presented his ideas about the brain's information-processing properties to a group of sensory

---

[*] If, while reading the last chapter, you wondered why it is that neurons in the visual system detect lines, information theory has an answer to that, too. In 1996, Bruno Olshausen and David Field applied a similar technique as Lewicki to show that lines are what you'd expect neurons to respond to if they are encoding images efficiently.

researchers gathered at MIT. When the proceedings of this meeting were translated into Russian for a Soviet audience, Barlow's contribution was conspicuously cut out. The Soviets, it turned out, had a problem with the use of information theory to understand the brain. Considered part of the 'bourgeois pseudoscience' of cybernetics, it ran counter to official Soviet philosophy by equating man with machine. Soviet leaders – and the sometimes scared scientists under their rule – openly critiqued this attitude as a foolish product of American capitalism.

Though unique in its political motivations, the Soviets' denunciation was far from the only critique of information theory in biology. In 1956, a short article entitled 'The bandwagon' cautioned against the overexcited application of information theory in fields such as psychology, linguistics, economics and biology. 'Seldom do more than a few of nature's secrets give way at one time. It will be all too easy for our somewhat artificial prosperity to collapse overnight when it is realised that the use of a few exciting words like *information*, *entropy*, *redundancy*, do not solve all our problems.' The article was written by Shannon himself, just eight years after he unleashed information theory on to the world.

Concerns about how apt the analogy is between Shannon's framework and the brain have even come from the very scientists doing the analogising. In a 2000 article, Barlow warned that 'the brain uses information in different ways from those common in communication engineering'. And Perkel and Bullock, in their original report, made a point of not committing themselves fully to Shannon's definition of information but, rather, treating the concept of 'coding' in the brain as a metaphor that may have varying degrees of usefulness.

The caution is warranted. A particularly tricky part of Shannon's system to map to the brain is the decoder. In a simple communication system, the receiver gets the encoded message through the channel and simply reverses the process of encoding in order to decode it. The recipient of a telegraph message, for example, would use the same reference table as the sender to know how to map dots and dashes back to letters. The system in the brain, however, is unlikely to be so symmetric. This is because the only 'decoders' in the brain are other neurons, and what they do with the signal they receive can be anyone's guess.

Take, for example, encoding in the retina. When a photon of light is detected, some of the cells in the retina (the 'on' cells) encode this through an increase in their firing rate while another set of cells (the 'off' cells) encodes it by decreasing their firing. If this joint up–down change in firing is the symbol the retina has designated to indicate the arrival of a photon, we may assume this is also the symbol that is 'decoded' by later brain areas. However, this does not seem to be the case.

In 2019, a team of researchers from Finland genetically modified the cells in a mouse's retina. Specifically, they made the 'on' cells less sensitive to photons. Now, when a photon hit, the 'off' cells would still decrease their firing, but the 'on' cells may or may not increase theirs. The question was: which set of cells would the brain listen to? The information about the photon would be there for the taking if the 'off' cells were decoded. Yet the animals didn't seem to use it. By assessing the animal's ability to detect faint lights, it appeared that the brain was reading out the activity of the 'on' cells alone. If those cells didn't signal that a photon was detected, the animal didn't respond. The scientists took this to mean that the brain is not, at least in this case, decoding all of the encoded information. It ignores the signals the 'off' cells are sending. Therefore, the authors wrote, 'at the sensitivity limit of vision, the decoding principles of the brain do not produce the optimal solution predicted by information theory'. Just because scientists can spot a signal in the spikes doesn't mean it has meaning to the brain.

There are many reasons this might be. One important one is that the brain is an information-*processing* machine. That is, it does not aim to merely reproduce messages sent along it, but rather to transform them into action for the animal. It is performing computations on information, not just communicating it. Expectations about how the brain works based solely on Shannon's communication system therefore miss this crucial purpose. The finding that the brain is not optimally transmitting information does not necessarily indicate a flaw in its design. It was just designed for something else.

Information theory, invented as a language for engineered communication systems, couldn't be expected to translate perfectly to the nervous system. The brain is not a mere telephone line. Yet parts of the brain do engage in this more basic task of communication. Nerves do send signals. And they do so through some kind of code based on spike rates or spike times or spike something. To glance at the brain from the vantage point of information theory, then, is a sensible endeavour – one that has yielded many insights and ideas. Stare too long, though, and the cracks in the analogy become visible. This is the reason for wariness. As a metaphor, the relationship between a communication system and the brain is thus most fruitful when not overextended.

CHAPTER EIGHT

Movement in Low Dimensions

Kinetics, kinematics and dimensionality reduction

In the mid-1990s, a local newspaper editor in Houston, Texas, went to Baylor College of Medicine hoping to get help with a problem concerning his left hand. For the past several weeks, the fingers of this hand had been weak and the fingertips had gone numb. The man, a heavy smoker and drinker, appeared to the doctors to be in decent health otherwise. Seeing the numbness, the doctors initially searched for a pinched nerve in his wrist. Finding nothing there, they checked the spinal cord, suspecting that a lesion in the spinal nerves might be to blame. When evidence for that too came back negative, the doctors went one step further and performed a scan of the brain. What they found was a tumour – the size of a large grape – lodged into the right side of the wrinkled surface of the man's brain. It was halfway between his right temple and the top of his head, in the middle of a region known as the motor cortex.

The motor cortex takes the shape of two thin strips starting at the top of the head and running down each side, together forming a headband across the top of the brain.* Different parts of each strip control different parts of the opposite side of the body. In the case of the newspaper editor, his tumour was in the hand-controlling region of the right motor cortex. It also extended a bit into the sensory cortex – a similarly arranged strip right behind the motor cortex that controls sensation. This placement explained the weakness and numbness respectively – problems that subsided after the surgical removal of the tumour.

Since its discovery roughly 150 years ago, the motor cortex has found itself at the centre of many controversies. That the brain controls the body is undisputed; data from injuries indicated this fact as early as the pyramid age of ancient Egypt. But *how* it does so is another question.

In some ways, the connection between the motor cortex and movement is straightforward. It's a connection that follows a path opposite to the investigation taken by the Baylor doctors: neurons in the motor cortex on one side of the brain send outputs to neurons in the spinal cord on the other side, and these spinal cord neurons go directly to specific muscle fibres. The spot where the spinal cord neuron meets the muscle is called the neuromuscular junction. When this neuron fires, it releases the neurotransmitter acetylcholine into this junction. Muscle fibres respond to acetylcholine by

---

* Technically this describes the 'primary' motor cortex. Another region known as the 'premotor cortex' sits just in front of the primary motor cortex. These areas are frequently studied together and we will not attempt to differentiate between them in this chapter.

contracting and movement occurs. Through this path, neurons in the cortex can directly control muscles.

But this is not the only road between the motor cortex and muscle. The other paths are more meandering. Some motor cortex neurons, for example, send their outputs to intermediate areas such as the brainstem, basal ganglia and cerebellum. From these areas the connections then go on to the spinal cord. Each such pit stop offers an opportunity for further processing of the signal, causing a change in the message that gets sent to the muscles. Furthermore, even the most direct paths aren't necessarily simple; neurons from the motor cortex can connect with multiple different neurons in the spinal cord, which each activate and inhibit different muscle groups. In this way, there are many channels through which the cortex can communicate with the muscles and many possible messages that could be sent. Rather than direct, the influence of the motor cortex on the body can actually be highly distributed.

On top of this confusion, the very need for the motor cortex has been called into question. When the cortex is disconnected from the rest of the brain, animals don't initiate many complex movements on their own, but they can still perform some well-worn responsive behaviours. For example, such 'decorticated' cats will claw and strike if restrained, and decorticated male rats still manage to copulate if a female is around. Therefore, for some of the most important behaviours for survival, the motor cortex seems superfluous.

Movement – as the only way the brain can communicate with the world – is a crucial piece of the neuroscience puzzle. Yet, the precise purpose of the motor cortex is debated and its anatomy is of little help in understanding it either. Without these inroads, it's hard to know exactly what the motor cortex is trying to say. But given the many significant motivations to sort out the riddle of movement – curing motor diseases, creating human-like robots, *etc.* – a steady stream of scientists has been trying. In the early days, this took the form of bitter arguments over what kind of movement the motor cortex generates. This was followed by a march of mathematical methods for making sense of the activity of its neurons. While some of its wilder debates have been brought under control, the study of the motor cortex – perhaps more than most areas of neuroscience – remains in tumultuous times even today.

★ ★ ★

Gustav Fritsch and Eduard Hitzig both studied medicine at the University of Berlin in the mid-nineteenth century, though their paths did not cross there. After medical school – the story goes – Fritsch was dressing a head wound as part of his service in the second Danish–Prussian War when he realised that certain irritations of the exposed brain caused muscle spasms in the opposite side of the soldier's body. Hitzig, on the other hand, was purportedly operating an electroshock-therapy business when he noticed that shocks to particular parts of the head elicited eye movements. Each doctor was struck by his own curious observation and the implications it could have. Fritsch and Hitzig finally met when Fritsch returned to Berlin in the late 1860s. The pair decided to join forces to explore what was then considered an absurd hypothesis: that the cortex could control movement.

The notion that the cortex did anything at all was considered radical at the time. Cortex – from the latin for 'rind' – was thought of as an inert outer shell, a layer of neural tissue covering the important areas of the brain underneath. This view came from several previous experiments that attempted to stimulate the cortex, but failed to elicit any interesting responses (in hindsight it is clear that this was due mostly to inappropriate stimulation techniques, such as pinching, poking or dousing the cortex with alcohol). But Fritsch, an international traveller who dabbled in many fields,* and Hitzig, a stern, proud and vain man, were driven by their own curiosity and arrogance to push through the barrier of established beliefs.

So, on a table in Hitzig's home (the Physiological Institute didn't have the facilities for this new technique), Fritsch and Hitzig began electrically stimulating the cortex of dogs. They prepared a platinum electrode by attaching it to a battery and tapping it on their tongue to test the strength (reportedly it was 'of sufficient intensity to cause a distinct sensation'). They then touched the tip of the electrode very briefly to different areas of the exposed brain, while observing any movements that came as a result. What they found was that stimulating the cortex could indeed produce movements – short twitches or spasms of small groups of muscles on the

---

* Including what we would now call 'scientific racism'. Fritsch collected samples of retina and hair in a vain attempt to make statements about white racial superiority.

opposite side of the body. And the location of the stimulation mattered – it determined which body part, if any, would move.

This latter finding was perhaps even more heretical than the first. At this time even the handful of scientists who thought the cortex might do something useful still assumed that it worked as an undifferentiated mass – a web of tissue without functional specialisation. It wasn't supposed to have an orderly arrangement, wherein all the motor functions sit at one strip across the front of the brain. Yet this is what Fritsch and Hitzig's stimulations were suggesting. Further testing their theory, after Fritsch and Hitzig mapped out the bit of the brain responsible for a certain body area, they would cut it out and observe the adverse effects this had on movement. In general, such lesions did not cause complete paralysis of the affected body part, but they did significantly impair its control and function. The evidence for a motor cortex was mounting.

The work of this German duo coincided with findings by another doctor, John Hughlings Jackson, that also implicated this region of the brain in motor control in humans. These joint findings made the mid-nineteenth century a turning point for the role of the cortex in movement – and in neuroscience more generally. Scientists were forced to contend with the idea that not only might the cortex be doing something, but that sub-portions of the cortex might even have different functions. Excited by these turbulent times, a mentee of Jackson's named David Ferrier set out to study the cortex in detail.

In 1873, Ferrier was given the opportunity to carry out experiments on the motor system at the West Riding

Lunatic Asylum. Working at this prestigious Victorian mental institution and research facility, Ferrier was able to replicate the findings of Fritsch and Hitzig – both the stimulation and the lesion results – in dogs. He also showed that the same principles applied in a whole host of other animals including jackals, cats, rabbits, rats, fish and frogs. Ferrier then went on a detailed expedition into the motor area of monkeys, hoping to provide a map that would help surgeons safely remove tumours and blood clots from human brains.*

Testing a zoo's worth of species wasn't the only way Ferrier expanded on Fritsch and Hitzig's work: he also updated the stimulation technology. The galvanic stimulation used by Fritsch and Hitzig was a form of direct current, which could be damaging to the tissue of the brain. As such, it could only be applied in brief pulses. Ferrier eventually switched to faradic stimulation, an alternating current that could be applied more consistently. As a result, Ferrier could hold the electrode to the brain for several seconds. He could also stimulate at higher strengths – 'sufficient to cause a pungent, but quite bearable, sensation when the electrodes were placed on the tip of the tongue', according to Ferrier.

---

* Like many scientists of his time, Ferrier engaged in animal experimentation techniques that would never pass ethical review today. Unlike many scientists of his time, however, Ferrier actually got in trouble for it. After Ferrier brought a lesioned animal on stage to demonstrate his findings in 1881, anti-vivisectionists had him arrested for performing an experiment without a licence. As it turned out, his associate, who did have a licence, was the one who had performed the surgery and Ferrier got off. But it was a strong sign from the animal rights crowd.

This quantitative change in stimulation parameters led to a qualitative shift in results. When Ferrier stimulated at longer intervals, he didn't just get longer muscle twitches. Instead, the animals displayed full, complex movements – movements reminiscent of ones they would make in their ordinary lives. For example, Ferrier found a part of the rabbit brain that, according to his notes, elicited 'sudden retraction and elevation or pricking up of the opposite ear – this occasionally coinciding with a sudden start, apparently as if the animal were about to bound forward'. In the cat, a particular area was responsible for 'retraction and adduction of the fore-leg. Rapidly performed, the movement is such as the cat makes when striking a ball with its paw.' And when stimulating a spot 'situated on the posterior half of the superior and middle frontal convolutions' of the monkey brain, Ferrier observed 'the eyes open widely, the pupils dilate, and head and eyes turn towards the opposite side'.

That such a blunt stimulation of the motor cortex conjures such smooth and coordinated movements points to a different understanding of this brain region from Fritsch and Hitzig's. If motor cortex stimulation primarily drives small isolated muscle groups – as those two believed – then its duty is relatively rudimentary. Different portions of motor cortex work like keys on a piano, each producing its own single note. Ferrier's findings, on the other hand, cast the motor cortex as a library of short melodies – where each stimulation causes a fragment of movement to play out cooperatively across multiple muscle groups. This argument – notes versus melodies, twitches versus movements – would be the first of many debates over the soul of the motor cortex.

Despite the fact that he replicated their findings, Fritsch and Hitzig didn't get along with Ferrier. This may have stemmed from a tiff over credit assignment. Ferrier felt that the German duo had slighted his mentor Jackson by not citing Jackson's work in their own. As a result, Ferrier tried to get away with not citing Fitsch and Hitzig in his. He even went so far as to remove all references to his experiments in dogs and only talk about the results from monkeys in order to avoid any connection to Fitch and Hitzig.[*]

Whatever the reason, Fritsch and Hitzig didn't trust Ferrier's work showing the production of natural movement fragments. They stood by the superiority of short galvanic stimulation, and claimed that Ferrier's stimulation was too long and that his findings were not replicable. Ferrier, on the other hand, remained steadfast in his belief that faradic current was best and that the quick galvanic application of Fritsch and Hitzig 'fails to call forth the definite purposive combination of muscular contractions, which is the very essence of the reaction and the key to its interpretation'.

As this fight raged on in the tiny microcosm of motor cortex stimulation, the larger field of neuroscience was having a different discussion. That function in the cortex is localised – that different areas play different roles – was something the motor crowd had clearly accepted. But in

---

[*] Over time, however, Ferrier seemed to cool. In his book published three years after the citation fight, he wrote: 'The credit of having first experimentally demonstrated the fact of definite localisation belongs to [Hitzig] and his colleague Fritsch, and I regret that in the acrimonious discussions which have arisen in regard to the subject, I have been interpreted as implying otherwise.'

the broader community, this radical change in understanding was still only just sinking in and many researchers at the time made it their goal to test the limits of this theory. Thus, the fashion became to try to stimulate as small an area of the cortex as possible to see just how localised the function could get. This trend aligned very well with Fritsch and Hitzig's approach of using short small pulses to cause individual muscle movements. Their preferred method thus became dominant – not because it was the right way to answer the scientific question of what the motor cortex does, but because the scientific question had changed. Under the influence of the localisation craze, the fact that small muscle movements could be elicited by stimulation was more important than whether the brain normally produces movement this way. The problem of 'twitches versus movements' was thus put on the back burner for more than a century.

★ ★ ★

In the brain, it doesn't get more local than a single neuron. Neuroscientists have had the ability to record the activity of single neurons since the late 1920s. However, doing so usually required an experimental set-up – say, removing the neural tissue from the animals or at least anesthetising them during the recording – that prohibited the concurrent study of behaviour. This changed in the late 1950s, when electrodes were developed that could be dropped into the brain of an awake and responsive monkey to eavesdrop on the electrical signal of individual neurons. Reflecting on this turn in the history of neuroscience some 50 years later, pioneering neuroscientist Vernon Mountcastle remarked that the field 'has never been the same since, and one can scarcely exaggerate the thrill it was for those of us who had spent years working with anesthetised or reduced preparations to see and to work with the brain in action!' The study of motor control – of the way in which movement and behaviour is generated by the brain – perhaps had the most to gain from this experimental advance. And indeed motor scientists were the first to use it.

One of these early researchers was Edward Evarts, a psychiatrist from New York. Evarts was a generous but strict man. He took his work very personally – even turning to introspection and personal experience to aid his scientific studies on sleep and exercise – and he expected similar dedication from others. In 1967, while working at the National Institutes of Health in Bethesda, Maryland, he finished a solo three-part project on the responses of neurons in the motor cortex. The final portion of this study focused on a question that would form the core of motor neuroscience for decades to come: what aspects of movement do neurons in the motor cortex represent?

To ask this question, Evarts used a simple motor task that required only a small amount of movement. Specifically, he trained monkeys to hold on to a vertical rod and move it left and right. The monkeys were constrained to do so using only a single joint, the wrist. This meant the movement was controlled solely by two muscle groups in the forearm: the flexors, which bring the hand towards the body, and extensors, which pull it back the other way.

One simple hypothesis was that the firing rates of neurons in the motor cortex were directly related to the position the wrist was in at any given moment. If this were the case, you'd find neurons that fire strongly when the wrist is flexed but not when it is extended and other neurons that do the opposite.

When studying movement, it makes sense to turn to the well-established mathematics of motion. An additional hypothesis for Evarts' study therefore came from *kinetics*, the part of physics that deals with the causes of motion.* To make the hand move, muscles in the arm contract, generating force. This force from the muscles gets turned into angular force, or torque, when it acts on the wrist joint. The torque then determines the movement and position of the hand. If motor cortex neurons encoded force instead of position, some neurons would fire strongly when the flexor muscles were producing force that moved the wrist in one direction and other neurons would fire when the extensor muscles were producing force that moves it in the other direction.

In the most basic form of this experiment, these two hypotheses are indistinguishable. If a neuron is firing when the wrist is in the flexed position, is it because the wrist is in the flexed position or because of the force needed to keep it there? Who knows? If he wanted to test between these two hypotheses, Evarts needed to dissociate these two aspects of movement. To do so, he simply added counterweights to the rod. Like setting the weight on a machine at the gym, adding different counterweights to the rod makes the movements easier

---

* Also commonly referred to as 'dynamics'.

or harder. This changes the amount of force needed to move the rod to the same position. Now firing rates could be compared when the wrist is in the same position but using different forces to get there.

When looking at the 31 neurons in the motor cortex from which he recorded, Evarts saw that 26 of them had firing rates that were clearly related to force. Some of these responded strongly during wrist flexion and increased their firing as weights were added that made that flexion more difficult (or decreased their firing when weight was added in the opposite direction, making flexion easier). Other neurons preferred extensions, showing the same pattern but in the opposite direction. The remaining five neurons were difficult to interpret, but none of them showed activity that related directly to wrist position. These results made a strong case for the argument that the motor cortex encodes force.

Evarts' work on wrist forces set in motion a long trajectory of searching for kinetics in the motor system. Over the ensuing years, several other research groups sought and found kinetic information in the firing rates of motor cortex neurons as animals carried out simple movements. The legacy of localisation remains visible in this approach; it aims, after all, to understand the behaviour of individual neurons and small isolated muscle movements. But it unifies that understanding under the broader, pre-existing mathematical system of kinetics. In this view, the mathematics of the motor cortex can be found in the equations of any standard physics textbook.

Evarts established the modern approach to the study of the motor cortex. He provided a well-controlled experiment to explore how the activity of an individual neuron relates to the activity of muscles, and he pointed to the mathematics that could contextualise such findings. Within a few decades, however, the majority of Evarts' signature contributions to motor science would be upended and the next era of tumult in the field would begin.

★ ★ ★

Ah, the motor system! For better or worse, there is no coherent view of motor function by systems neuroscientists ... You still have the muscle diehards [who] contend that all neural activity in motor structures, from the cerebellum to cortex, has somehow to be explained, by default, with reference to that one, real or virtual, muscle. Of course, it doesn't make any sense... natural movements rarely, if ever, involve just one muscle.

These are the words of Apostolos Georgopoulos, a Greek-born professor of neuroscience at Johns Hopkins University. By the time he said this in 1998, Georgopoulos had been shaking up the field of motor neuroscience for more than 15 years. Three major conceptual advances are associated with Georgopoulos (though, of course, none of them came solely from him – they existed in some form already in the scientific milieu) and of those three contributions, two remain central to the study of the motor cortex to this day.

His first contribution is predictable from his quote: a focus on natural movements. Georgopoulos had trained with the famed sensory neuroscientist Vernon Mount-castle and was heavily influenced by his way of thinking.

Mountcastle took a holistic approach to studying the brain. He asked how bodily sensations were represented at every step – from tactile-sensing neurons in the skin up through to the use of this sensory information by higher cognitive functions in the brain. Georgopoulos wanted to bring the study of motor control into the grand tradition set upon by Mountcastle in sensory systems. In this pursuit, Georgopoulos knew he had to do away with studies of stilted single-joint movements. To understand how movement information was represented and processed by the brain, he needed to study it in the context of natural movements, in their full multi-muscle complexity. For this, he turned to one of the most basic and essential movements in a primate's repertoire: reaching.

To reach for an object in front of you, you rely on the team of muscles that surround the joints of the arm. This unsurprisingly includes the muscles of the upper arms such as the biceps and triceps. But it also involves the anterior pectoralis (a band that crosses from the centre of the chest to the arm), the anterior deltoid (a strip just in front of the armpit) and the broadest muscle in the back, the latissimus dorsi, which stretches from the lower back up to the under arm. Depending on the details of the reach, the wrist and fingers may need to be involved as well. This multi-muscle movement is a far cry from the wrist-bend experiments of Evarts.

To study this multifaceted task, Georgopoulos trained monkeys to work on a small light-up table. The animals held in their hand a rod, just like you'd hold a wooden spoon when stirring a big pot. This rod was connected to a measurement device and, when a light turned on that indicated where to reach, the animals moved the rod to that location. The lights were arranged in a circle, like the numbers on the face of a clock, with a radius about the length of a playing card. The monkeys always returned to the centre of the circle before reaching to the next location. With eight evenly spaced lights in the circle, the animals thus made movements in eight different directions. This simple set-up became part of a tradition of studying 'centre-out' reaching in motor neuroscience (see Figure 18).

For his second contribution, Georgopoulos targeted Evarts' kinetics theory – and replaced it with his own take on what neurons in the motor cortex were representing.

Evarts had used wrist-flicks to see that neural activity represented force, but some scientists had found this relationship inconsistent. During more complex movements in particular, the amount of force that a muscle produces changes as the joints and muscles around it do. Moving the shoulder, for example, changes the physics of the elbow. This makes the chain of influence that runs from neural activity to muscle activity to force less interpretable and the kinetic theory

'Direction tuning' in the reach task

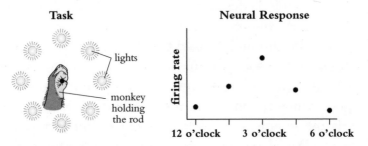

Figure 18

less viable. There were also indications that plenty of neurons cared very little about force at all.

So Georgopoulos flipped the perspective. Rather than asking what neurons were saying about the muscles, he asked what they were saying about the movement. In more than a third of the neurons in motor cortex, he found a very clear and simple relationship between neural activity and the *direction* the arm was moving. Specifically, these neurons had a preferred direction. This means they would fire the most when the animal reached in that direction – say towards three o'clock – and their rate of firing would fall off the farther the movement was from this direction (lower firing for two and four o'clock, lower still for one and five o'clock, *etc.*). The finding of this 'direction tuning' implied that the motor cortex cared more about *kinematics* than kinetics. After 'twitches vs movements', kinetics vs kinematics became the next big debate in motor neuroscience.

Kinematics are descriptive features of motion, defined without regard to the forces that generate them. In this way, kinematic variables indicate the desired outcome for the arm, but not the instructions for how to create it. Going from a model wherein the motor cortex encodes kinetics to one wherein it encodes kinematics shifts the distribution of labour in the motor system. A kinetic variable such as force is only a few small calculations removed from the actual level of muscle activity needed to enact it – transformations potentially performed by neurons in the spinal cord – but because kinematic values only define where the arm should be in space, they pose a larger challenge to the rest of the motor system. The burden is thus on areas downstream of the motor cortex to change the coordinate system: to take a set of desired locations external to the body and turn it into patterns of muscle activity. Georgopoulos became an adamant and unceasing defender of this kinematic take on the motor cortex for decades to come.

The final change ushered in by Georgopoulos was in how to look at the data. If neurons are tuned to the overall direction of the movement, it means that they are not involved in one-to-one control of muscles. Why, then, should neurons be looked at one at a time? Instead, it would be more sensible to consider what all the neurons – the entire population – are saying.

And that is what Georgopoulos did. Using the information he had about direction tuning in individual cells, he calculated a 'population vector' – essentially an arrow that points in the direction of movement that the neural population is encoding. This calculation works by allowing each neuron to vote for its preferred direction of movement. But this is no perfect democracy, because not all votes are weighted equally. Instead the weight of a neuron's vote comes from how active the neuron is. So, neurons firing high above average can strongly bias the population vector in their preferred direction. And neurons firing well below average would bias this vector *away* from their preferred direction. In this way, the neurons collectively indicate the desired movement direction. And they do so in a way that is more accurate than an individual neuron can be.* Georgopoulos

---

* To understand this, consider the previously described neuron that prefers movement towards three o'clock. That neuron has a unique firing rate for three o'clock, but may fire at the same rate for

showed that by summing up each neuron's contribution in this way he could, in fact, accurately read out the direction in which the animal was moving its arm on any given instance.

This population-level approach to the data proved quite powerful – perhaps too powerful. After this work, several studies emerged showing other information that could be read out from the motor cortex if the whole population was taken into account: finger movement, arm speed, muscle activity, force, position and even sensory information about the visual cues indicating where and when to move. Direction may have been one of the original variables decoded in this way, but it was far from unique. Finding both kinetic and kinematic variables (and a host of other information) in the activity was a blow against Georgopoulos' theory that kinematics is special. Ironically, it was one of his own contributions – a focus on the population – that undermined him.

Faith in what information read out from the motor cortex tells us about its function declined even further through the use of computational approaches. Scientists who built models of the motor system, which they *designed* to work by encoding kinetic variables, were able to show how kinematic variables could be read out from them as well. One such scientist – Eberhard Fetz – went so far as to compare the hunt for variable representations in the motor cortex to the behaviour of fortune tellers:

---

movements towards two and four o'clock, making those directions indistinguishable. When combining information from that neuron with others that have different preferred directions, however, that ambiguity can be resolved.

'Like reading tea leaves, this approach can be used to create an impression, by projecting conceptual schemes on to suggestive patterns.'

These findings revealed a possibility that was always lurking just under the surface: there is no single value that neurons in the motor cortex 'code for'. It's not kinematics *or* kinetics; it's both, and more, and neither. In many ways, this fact was visible all along. It could be seen in the neurons that didn't show neat responses to force or direction, or in the neurons that showed big changes in their responses with small changes in the experiment, or simply in the decades-long volley of researchers finding evidence for one side and then the other over and over again.

By some accounts, the field was led astray because it blindly followed the path laid by other scientists: the study of sensory systems that so inspired Georgopoulos was a poor model for how to understand movement. The debate over 'what does the motor cortex encode?' was unresolved not because the question is hard, but because it was the wrong one to ask from the start. The motor system doesn't need to track movement parameters, it just needs to produce movements.

As was shown in the last chapter, just because scientists can see structure in neural activity doesn't mean the brain is using it. A common analogy compares the motor cortex to the engine of a car. The engine is surely responsible for the movement of the car. And if one were to measure the activity of its different parts – the piston, the engine belts, *etc.* – it's likely that some of these values would, under some conditions, correlate quite strongly with the force generated by the car or with the direction in which it is turning. But would we say the engine works by 'encoding' these variables? Or is this more a convention brought to the motor cortex by scientists who themselves understand movement through concepts of force, mechanics and physics? As motor neuroscientist John Kalaska wrote in 2009: 'Joint torque is a Newtonian mechanical parameter that defines the rotational force required to produce a particular joint motion ... It is highly unlikely that a [motor cortex] neuron knows what a Newton-meter is or how to calculate how many Newton-meters are needed to generate a particular movement.'

★ ★ ★

Just because 'what does the motor cortex encode?' was the wrong question to ask to understand the motor cortex doesn't mean the answer has no value. Indeed, trying to decode information from the motor cortex can be quite useful – not in aid of understanding the motor system, but in bypassing it altogether.

When, in a small room outside Providence, Rhode Island, a 55-year-old woman named Cathy Hutchinson brought a cup of coffee to her mouth and took a sip, it was the first time she'd done so in more than 15 years. It was also the first time any tetraplegic person achieved this feat. Hutchinson became paralysed from the neck down when she was 39, after suffering a stroke while gardening on a spring day in 1996. According to a *Wired* interview, she heard about BrainGate – a research group housed at Brown University that explores the use of brain–computer interfaces to restore mobility to patients – from a friend who worked at the hospital. Hutchinson enrolled in their clinical trial.

As part of their study, the BrainGate scientists implanted a device in Cathy's brain – a square piece of metal smaller than the average shirt button, consisting of 96 electrodes, placed in the arm-controlling region of her left motor cortex. The activities of neurons recorded through these electrodes are fed through a wire out of her head and into a computer system. This machine connects to a robotic arm placed on a stand to Cathy's right. The arm itself is alien – bulbous, awkward and a shiny blue – but the hand at the end of it is more recognisable, with subtle detailed joints in a matte silver. When Cathy controls it, the movements are not smooth. The arm halts and moves backwards before eventually bringing the coffee towards her. But ultimately it gets the job done – and a woman who lost the ability to move her own limbs gained the ability to move this one.

Even this simple, imperfect control did not come immediately. For the machine to know how to listen to Cathy's motor cortex it had to be trained. BrainGate achieves this by having Cathy imagine moving her arm in different directions. The patterns in the neural activity can then be related to commands to move the arm in different directions. In this way, the control of this brain-computer interface depends on the presence of direction tuning in the motor cortex. That is, if it weren't possible to read out movement direction – and other intentions such as grasp or release – from a population of motor cortex cells, brain-computer interfaces wouldn't work.

These devices also depend on a lot of heavy mathematical machinery that runs behind the scenes. To try to make the movements as smooth as possible, for example, BrainGate uses an algorithm that combines input from the neural activity with information about past movement direction. Furthermore, some of the fine-grained movements of the robot hand and wrist are pre-programmed into the device, so that the user can initiate a full, detailed motor sequence by imagining a simple command. This is a practical workaround for how difficult it is to read out detailed motor commands from the activity of a group of neurons.

Such laboured efforts to build brain-controlled motor devices, while offering some hope to patients, highlight just how little we understand the role the motor cortex plays in healthy movement.

★ ★ ★

If decoding is more useful for engineering than understanding, what *can* we use to make sense of the motor cortex?

Focusing on the population has remained popular – and with good reason. The neurons of the motor cortex must to some extent be working together to generate movements; after all, the human motor cortex is allotted hundreds of millions of neurons to control the roughly 800 muscles of the human body. But this poses a challenge for scientists: how can they make sense of the activity of the hundreds of neurons normally recorded in the course of an experiment? In the population vector approach taken by Georgopoulos, beliefs about what the neurons were doing was built in: if the neurons are encoding direction, then direction is what is read out.

But as motor neuroscientists tried to move away from the question of what the motor cortex encodes to what the motor cortex *does*, approaches based on reading out specific information didn't make sense. They needed to find a new way to look at the population.

The fundamental difference between a single neuron vs a population approach to the study of the motor cortex is one of *dimensionality*. While the space we live in is three-dimensional, many of the systems scientists study have much higher dimensionality. The activity of a population of 100 neurons, for example, would be 100-dimensional.

How this abstract, high-dimensional 'neural space' relates at all to real, tangible, physical space can be difficult to see. But we can build off our intuitions about physical space by first considering a population with only three neurons. Specifically, by replacing metres or feet with the number of spikes a neuron fires, the activity of this population can be described just like a location in space. When producing a movement, for example, the first neuron in the population may fire five spikes, the second fifteen, and the third nine. This provides coordinates in neural space, the same way a treasure map describes how many paces to walk forwards, then to the right, then how deep to dig. A different pattern of neural activity would point to a different location in neural space. By looking at the neural activity of the motor cortex across a variety of movements, scientists can ask if different locations in this space correspond to different types or components of movement.

Visualising this activity, though, becomes tricky in a larger population. Humans, based as we are in our three-dimensional world, have trouble thinking beyond it. What would neural space look like if a fourth neuron were added to the population? What if there were a hundred neurons, or a thousand? Our intuitions fall apart here. Computer scientist Geoffrey Hinton offers the best advice for this problem: 'To deal with hyper-planes in a fourteen-dimensional space, visualise a 3-D space and say "fourteen" to yourself very loudly.'

Luckily there is another way around the problem of having too many dimensions: dimensionality reduction. Dimensionality reduction is a mathematical technique that can take information in a high-dimensional space and represent it using fewer dimensions. It is based on the premise that some of these original dimensions are redundant – in this case, that would mean that multiple neurons are saying the same thing. If you could figure out what patterns of neural activity in a 100-dimension population are fundamental to that population – and which are just recycled combinations of those fundamental patterns – you could explain that neural population with fewer than 100 dimensions.

Consider personality. How many dimensions are there to human personality? The English language has a dizzying list of possible descriptors: agreeable, flexible, self-critical, kind, forgiving, creative, charismatic, calm, intelligent, disciplined, aggressive, meticulous, serious, clever, and on and on and on. Each of these could be thought of as a separate dimension, with each person described by a location in this high-dimensional personality space based on how they score on them. But there are some personality traits that seem to correlate. For example, 'clever' people may also frequently be considered 'quick-witted'. Maybe it would be more appropriate to think of cleverness and quick-wittedness as two measures of the same underlying trait – perhaps we call it 'intelligence'. If this is the case then the two dimensions representing cleverness and quick-wittedness in this space can be replaced by one for intelligence. This reduces the dimensionality. If there is only an occasional person who is clever but not quick-witted or quick-witted but not clever, then this reduction doesn't sacrifice much. For the vast majority of people, describing them according to intelligence alone will tell us what we need to know about these aspects of their personality.

Indeed, most popular personality tests are based on the premise that just a handful of core traits can explain the full human diversity. For example, the famed Myers-Briggs test claims personality has just four axes: intuition vs sensing, feeling vs thinking, introversion vs extraversion and perception vs judging. A more scientifically grounded approach (known as the Big Five) places the dimensionality of personality at five: agreeableness, neuroticism, extraversion, conscientiousness and openness. These are referred to as 'latent' factors, because they can be thought of as the basic underlying traits that generate the many different personality styles we see.

The historic tradition in neuroscience of treating each neuron as a snowflake – unique and worthy of individual analysis – assumes that they are, in some sense, the base unit of the brain. That is, it assumes that nature has packaged the relevant dimensions into a neat cellular form. But just as our folk notions of personality over-represent its dimensionality, there are many reasons why the 'true' dimensionality of a neural population is likely to be less than the number of neurons in it. For example, redundancy is a smart feature to have in any biological system. Neurons are noisy and can die, which makes a system with redundant ones more robust.* Furthermore, neurons tend to be heavily interconnected. With all their talking back and forth with each other, it's unlikely that any one of them can remain very independent. Instead, their activity becomes correlated the same way that the opinions of people in the same social circles start to converge. For these reasons, neural populations are ripe for applying dimensionality reduction techniques that can help identify the latent factors truly driving them.

A popular dimensionality reduction technique for neural data is principal components analysis, or PCA (see Figure 19). PCA was invented in the 1930s and used extensively by psychologists to analyse mental traits and faculties. Because of how useful it is for making sense of big datasets, it's now applied to all kinds of data across many fields.

PCA works though a focus on *variance*. Variance refers to how spread out different data points are. If, for example, over the course of three nights a person sleeps for 8 hours, 8 hours and 5 minutes, and 7 hours and 55 minutes, they would be a low-variance sleeper. A high-variance sleeper may also sleep for 8 hours on average but would spread it out very differently over three nights – say 6 hours, 10 hours and 8 hours.

---

* This is technically at odds with Barlow's notions of efficient and sparse coding from the last chapter. Practically speaking, the brain needs to balance efficient information transmission with the need to be robust and so should have some redundancy.

Dimensions with high variance are important because they can be quite informative. It's easier, for example, to read the emotional state of someone who is sometimes quiet and sometimes screaming than a stoic who always has the same straight face. Similarly, it is easier to categorise people by traits that vary a lot across people, rather than ones that everyone has in common. Recognising the importance of variance, the goal of PCA is to find new dimensions – which are a combination of the original dimensions the same way intelligence could be a combination of cleverness and quick-wittedness – that capture as much of the variance in the data as possible. This means that knowing where a data point falls according to these new dimensions will still tell us a lot about it, even if there are fewer of them.

For example, consider a two-neuron population whose activity we would like to describe with just a single number. Let's say we recorded the activity of these two neurons during different movements, so for each movement we have a pair of numbers representing the number of spikes from each. If we plot these pairs using the x-axis for one neuron and the y-axis for another, we

Principal Components Analysis

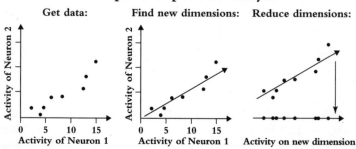

Figure 19

may see that the data falls more or less along a line. That line then can be our new dimension. Now, instead of describing the activity during each movement as a pair of numbers we can describe it as a single number, which refers to where it falls on this line.

Reducing the dimensionality in this way does leave off some information. We don't know, for example, how far the activity is from that line if we only describe where it falls on it – but the point is to pick the line that captures the most variance and thus loses the least information.

If the data doesn't fall along a line – that is, if the activity of the two neurons is not similar at all – then this won't work very well. In that case we would say that this two-dimensional neural population is indeed using its full two dimensions and so can't be reduced. But, as discussed previously, there are many reasons why, on average, some neural activity is redundant and so dimensionality reduction is possible.

Dimensionality reduction has been successfully applied to all kinds of neural data over the years. PCA itself was applied as early as 1978, when it was used to show that the activity of eight neurons responsible for encoding the position of the knee could be well represented by just one or two dimensions. And the use of PCA in studies of the motor cortex has only ramped up in the past decade. This is because dimensionality reduction helps motor scientists see what would otherwise be hidden. Condensing the ups and downs of the activity of over a hundred neurons into a single line makes their patterns visible to the naked eye. Looking at the evolution of a population's activity as a shape traced out in three dimensions allows scientists to use their intuitions about space to understand what neurons are doing. Seeing these trajectories can therefore seed new stories about how the motor system works.

Studies conducted in the lab of Krishna Shenoy at Stanford University in the early 2010s, for example, asked how the motor cortex *prepares* for movements. To do this, they trained monkeys to perform standard arm reaches, but they introduced a delay between when the location for the movement was given and when the animal should actually start moving. This made it possible to record from the motor cortex while it was preparing to move.

The long-held assumption was that, in preparing for movements, motor cortex neurons would fire in a pattern similar to how they fire during the movement, just at lower overall rates. That is, they'd essentially be saying the same thing but quieter. In neural activity space, this would put the preparatory activity in the same direction as the movement activity, but just not quite as far along. However, by plotting a low-dimensional version of the neural activity as the animal planned for and then executed a movement, the researchers found this was not the case. The activity before the movement was not simply a restrained version of that during the movement; instead, it occupied a different area of the activity space entirely.

This finding, while surprising, is consistent with a more modern view of the motor cortex. This new view puts emphasis on the fact that the motor cortex is a dynamical system – that the neurons within it interact in a way that makes them capable of producing complex patterns of activity over time. Because of these interactions between its neurons, the motor cortex has the ability to take in short, simple inputs, and produce elaborate and extended outputs in return. What makes this so useful is that it means another brain region could decide where the arm should be, send that information to the motor cortex, and the motor cortex would produce the full trajectory of neural activity needed to make the arm get there.

In this framework, the preparatory activity represents an 'initial state' of this dynamical system. Initial states define the location in activity space that the population starts at, but it is the connections among neurons that determine where it goes. In this way, initial states work a little like the entries to different water slides at the top of a platform: the location of the entry can have little to do with the course of the slide or where it ends up. Therefore, there is no reason preparatory activity needs to look like the activity during movement. All that matters is that the motor cortex gets to the right initial state and the connections among its own neurons will do the rest.

This 'dynamical systems' view has the potential to explain why attempts to make sense of the motor cortex have been so confounded. Thinking of these neurons as part of a larger machinery – where some parts are guiding muscle movements in the moment, but others are planning for the next step – the diversity and malleability of their responses are more expected. This new take also, ironically, brings the field back to its roots. A model where a simple input can produce a complex output provides a nice fit to Ferrier's findings that stimulation produces extended naturalistic movements. And Ferrier was indeed vindicated in the early 2000s, when Princeton professor Michael Graziano showed – using modern stimulation techniques – that half-second-long stimulation of the motor cortex elicits complex and coordinated natural movements, such as bringing a hand to the mouth or a changing of facial expression.

★ ★ ★

It is not uncommon for scientists to admit what they do not know. After all, science only exists in the gaps between knowledge, so identifying and recognising these gaps is an important part of the process. But researchers of the motor system seem particularly extreme in their declarations of ignorance. They fill paragraphs with talk of the 'considerable debate' in their field and how there is 'remarkably little agreement regarding even the basic response properties of the motor cortex'. They are quick to admit as well that 'a deep understanding of the motor cortex function still eludes us' and that 'it remains an open question how neural responses in the motor cortex relate to movement'. And in their more desperate moments they even ask: 'Why has this seemingly simple question been so difficult to answer?'

While still couched in the dry detachment of academic writing, these words indicate an honest acknowledgement of an unfortunate truth: that despite being both one of the earliest areas of cortex ever explored and one of the first to have single neuron activity recorded during behaviour, the motor cortex still remains deeply, stubbornly mysterious. As we have seen, of course, this is not for lack of trying: heroic work and vigorous debates have marked the history of this field – and, certainly, many inroads have been made. Yet, few of the major controversies have been fully settled – save, presumably, for the issue of whether the motor cortex exists and the fact that it does something at all.

CHAPTER NINE

From Structure to Function

Graph theory and network neuroscience

In 1931, three years before his death, Santiago Ramón y Cajal gave the Cajal Institute in Madrid a trove of his personal possessions. The collection contained all manner of scientific trinkets: balances, slides, cameras, letters, books, microscopes, solutions, reagents. But the items most notable – the ones that would become nearly synonymous with the name Cajal – were the 1,907 scientific drawings he had created throughout his career.

Most of these drawings were of different parts of the nervous system and were produced via a laborious cell-staining process. It started with a live animal, which was sacrificed and its tissues preserved. A chunk of the brain was then removed and soaked in a solution for two days, dried and soaked in a different solution – this one containing silver that would penetrate the cell structures – for another two days. At the end of this, the brain tissue was rinsed, dried again and cut into slices thin enough to fit on a microscope slide. Cajal looked at these slides through the eyepiece of his microscope and sketched what he saw. Starting first with pencil, he outlined every nook and cranny of each neuron's shape on a piece of cardboard, including the thick cell bodies and the thin appendages that emerged from them. He then darkened in the cells with India ink, occasionally using watercolours to add texture and dimension. The result was a set of haunting silhouettes of stark, black spider-like figures against beige and yellow backgrounds.* The exact contours and configurations depended on the animal and the nerve fibres in question; more than 50 different species and nearly 20 different parts of the nervous system are portrayed on Cajal's cardboard canvases.

These hundreds of portraits represent the infatuation Cajal had with the *structure* of the nervous system. He sought enlightenment in the basic unit of the brain – the neuron. He fixated on how they were shaped and how they were arranged. A focus on its physical foundations was Cajal's inroad to understanding how the brain worked. Function, he believed, could be found in structure.

And he was right. Cajal was able to deduce important facts about the workings of the brain by looking long and hard at how it was built. One of his significant findings was about how signals flow through neurons. Through his many observations of different neurons in different sensory organs, Cajal noticed that the cells were always arranged a certain way. The many branched dendrites of a cell would face the direction the signal was coming from. The long singular axon, on the other hand, went towards the brain. In the olfactory system, for example, neurons with the chemical receptors capable of picking up odour molecules exist in the mucousy skin

---

* The images were attractive enough to be the centrepiece of a travelling art exhibition called *The Beautiful Brain*, a fate that surely would've pleased Cajal. Before succumbing to his father's wishes that he be a physician, Cajal dreamt of being an artist.

inside the nose. These neurons send their axons up into the brain and make contact with the dendrites of cells in the olfactory bulb. These neurons then have axons that go further on into other parts of the brain.

This pattern – which Cajal saw over and over – strongly suggested that signals flow from dendrites through to axons. Dendrites, he concluded, act as the receiver of signals for a cell and axons the sender of signals on to the next cell. So clear was Cajal on this that he added little arrows to his drawings of circuits like the olfactory system, indicating the presumed direction of information flow. Cajal, as we now know, was exactly correct.

Cajal was one of the founding fathers of modern neuroscience. As such, his belief in the relationship between structure and function was entered into the DNA of the field. Reflections of this idea are peppered throughout neuroscience's history. In a 1989 article, Peter Getting wrote that researchers in the 1960s could see, even through their limited data, that 'the abilities of a network arose from the interconnection of simple elements into complex networks, thus, from connectivity emerged function'. Studies in the 1970s, he goes on to say, 'were approached with several expectations in mind: first, a knowledge of the connectivity would explain how neural networks operated'. This attitude persists. A review written in 2016 by professors Xiao-Jing Wang and Henry Kennedy ends with the statement: 'Establishing a firm link from structure to function is essential to understand complex neural dynamics.'

Structure exists at many scales in the brain. Neuroscientists can look at the shape of a neuron, as

Cajal did. Or they can look at how neurons are wired up: does neuron A connect to neuron B? They can zoom out even further and ask how small populations of neurons interact. Or they can investigate brain-wide connectivity patterns by looking at the thick bundles of axons that connect distant brain regions. Any of these higher-level structures may hold secrets about function as well.

But to unearth these secrets, neuroscientists need a way to see and study these structures clearly. Something that could've been considered a limitation of the staining method Cajal used – that it stained only a small number of neurons at a time – was actually an advantage that made it revolutionary. A method that stained all neurons in the field of view would've produced a black mess with no discernible structures; it would've missed the trees for the forest. Since neuroscientists have moved their study of structure away from single neurons and on to the more complicated subject of connections, networks and circuits, they may be at even higher risk of being overwhelmed by data and distracted by the wrong details.

A much-needed method, however, has been found in a particular subfield of mathematics: graph theory. The language of graph theory offers a way to talk about neural networks that cuts away much of the detail. At the same time, its tools find features of neural structure that are near impossible to see without it. These features of the structure, some scientists now believe, can inspire new thoughts on the function of the nervous system. Swept up by the promise of graph theory's methods, neuroscientists are currently applying it to everything from brain development to disease. Though the dust has not yet settled on this new approach to the brain, its fresh take on old problems is exciting many.

★ ★ ★

In the eighteenth-century East Prussian capital of Königsberg, a river branched in two as it cut through the town, creating a small island in the middle. Connecting this island with parts of the town north, south and east of it were seven bridges. At some point a question arose among the citizens of Königsberg: is there a way to navigate through the city that crosses each of the bridges once and only once? When this playful question met the famous mathematician Leonhard Euler, the field of graph theory was born.

Euler, a polymath who was born in Switzerland but lived in Russia, wrote 'Solutio problematis ad geometriam situs pertinentis' or 'The solution of a problem relating to the geometry of position' in 1736. In the paper, he answered the question definitively: a Königsberger could *not* take a walk through their town crossing each bridge exactly once. To prove this he had to simplify the town map into a skeleton of its full structure and work logically with it. He had shown, without using the word, how to turn data into a *graph* and how to perform computations on it (see Figure 20).

Within the context of graph theory, the word 'graph' does not refer to a chart or a plot, as it does in common language. Rather, a graph is a mathematical object, composed of *nodes* and *edges* (in the modern parlance). Nodes are the base units of the graph and edges represent the connections between them. In the Königsberg example, the bridges serve as edges that connect the four different land masses, the nodes. The degree of a node is the number of edges it has; the 'degree' of a land mass is thus the number of bridges that reach it.

Euler approached the bridge-crossing question by first noting that a path through town could be written down as a list of nodes. Giving each land mass a letter name, the list 'ADBC', for example, would represent a path that goes from the island at the centre to the land at the bottom (via any bridge that connects them), then from there to the land mass on the right and then on to the land at the top. In such a path through a graph, one edge is traversed between each pair of nodes. Therefore, the number of bridges crossed is equal to the number of letters in the list minus one. For example, if you've crossed two bridges, you'd have three land masses on your list.

Euler then noticed something important about the number of bridges that each land mass has. This number is related to how many times the land mass should show

Map of Königsberg   **As a graph**

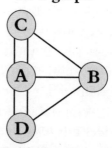

[   ] = river   [   ] = land   [bridge] = bridge

Each land mass is a node
given a letter name and
the bridges form the edges

Figure 20

up in the path list. For example, land mass B has three bridges, which means 'B' must appear twice in any path that crosses each bridge once – that is, there is no way to cross these three bridges without visiting B twice. The same is true for land masses C and D, as they both have three bridges too. And land mass A, with five bridges, needs to appear three times in the path list.

Taken together, any path that satisfies these requirements would be nine (2+2+2+3) letters long. A list of nine letters, however, represents a path that crosses *eight* bridges. Therefore, it is impossible to build a path that crosses each of the seven bridges only once.

Using this relationship between a node's degree and the number of times the node should appear in a path, Euler derived a set of general rules about what paths were possible. He could now say for *any* set of bridges connecting *any* patches of land whether a path crossing each bridge only once existed.

More than that, it doesn't matter if we are talking about land and bridges at all. The same procedure could be used to find paths for a town snowplough that needs to clear each street only once or to see if it's possible to traverse Wikipedia clicking each hyperlink between sites just one time. This pliability is part of what gives graph theory its potency. By stripping away the details of any specific situation, it finds the structure that is similar to them all. This abstract and alien way of looking at a problem can open it up to new and innovative solutions, just as treating a walk through town as a list of letters helped Euler.

Given this feature, graph theory found purpose in many fields. Chemists in the nineteenth century wrestled for some time with how to represent the structure of molecules. By the 1860s, a system was developed that is still in use today: atoms are drawn as letters and the bonds between them as lines. In 1877, English mathematician James Joseph Sylvester saw in this graphical representation of molecules a parallel to the work being done by descendants of Euler in mathematics. He published a paper drawing out the analogy and used, for the first time, the word 'graph' to refer to this form. Since then, graph theory has helped solve many problems in chemistry. One of its most common applications is finding isomers – sets of molecules that are each made up of the same type and number of atoms, but differ in how those atoms are arranged. Because graph theory provides a formal language for describing the structure of atoms in a molecule, it is also well suited for enumerating all the structures that are possible given a particular set of atoms. Algorithms that do this can aid in the design of drugs and other desirable compounds.

Like a chemical compound, the structure of the brain lends itself well to a graph. In the most basic mapping, neurons are the nodes and the connections between them are the edges. Alternatively, the nodes can be brain areas and the nerve tracts that connect them the edges. Whether working on the microscale of neurons or the macroscale of brain regions, putting the brain into the terms of graph theory exposes it to all the tools of analysis this field has developed. It is a way of formalising the informal quest that has always guided neuroscience. To speak of how structure births function first requires the ability to speak clearly about structure. Graph theory provides the language.

Of course, there are differences between the brain and a Prussian town or a chemical compound. The connections in the brain aren't always a two-way street like they are on a bridge or in a bond. One neuron can connect to another without it receiving a connection back. This unidirectional nature of neural connections is important for the way information flows through neural circuits. The most basic graph structures don't capture this, but by the late 1800s, the concept of *directed* graphs had been added to the arsenal of mathematical descriptors. In a directed graph, edges are arrows that only flow one way. The degree of a node in a directed graph is thus broken down into two categories: the in-degree (for example, how many connections a neuron receives) and the out-degree (how many connections it sends out to other neurons). A study done on neurons in the cortex of monkeys found these two types of degree to be roughly equal, meaning neurons give as much as they receive.

In 2018, mathematicians Katherine Morrison and Carina Curto built a model of a neural circuit with directed edges in order to answer a question not so dissimilar to the Königsberg bridge problem. Rather than determining what walks through town a certain set of bridges can support, they explored what sequence of neural firing a given circuit could produce. By bringing in tools from graph theory, Morrison and Curto figured out how to look at a structure of up to five model neurons and predict the order in which they will fire. Ordered patterns of neuronal firing are important for many of the brain's functions, including memory and navigation. This five-neuron model may only be a toy example, but it perfectly encapsulates the power promised by bringing graph theory into the study of the brain.

For real brain networks, however, a more 'global' perspective needs to be taken.

★ ★ ★

Over the course of a few months in the late 1960s, a stockbroker living in Sharon, Massachusetts, was given 16 brown folders from the proprietor of a local clothing store. Strange though this was, the folders weren't a surprise to the stockbroker. They were simply part of an unorthodox social experiment being run by the famous social psychologist Stanley Milgram. With this experiment, Milgram wanted to test just how big – or small – the world truly was.

The phrase 'it's a small world' is usually uttered when two strangers meet and serendipitously discover that they have a friend or relative in common. Milgram wanted to know just how often something like this could happen: what are the odds that two people chosen at random have a friend in common? Or a friend of a friend? Framed another way, if we could see the entire network of human connections – a graph where each node is a person and each edge a relationship – what would the average distance between people be? How many edges would we need to traverse to find a path between any two nodes?

In a bold attempt to answer this question, Milgram chose a target person (in this case, the Massachusetts stockbroker) and several starters: unrelated people in another part of the country (in this case, mostly Omaha,

Nebraska). The starters were given a package with a folder and information about the target person. The instructions were simple: if you know the target, give the folder to them; otherwise, send it on to a friend of yours who you think has a better shot of knowing them. The next person would be told to follow the same instructions and, hopefully, eventually the folder would end up with the target. The senders were also asked to write their name in a register that was sent along with the package, so that Milgram could trace the path the folder took.

Looking at a total of 44 folders that made it back to the stockbroker, Milgram found that the shortest path consisted of just two intermediate people and the longest had 10. The median was just five. Getting the folder through five people between the starter and the target involved six handoffs and thus the notion of 'six degrees of separation' – already posited by observant scientists and sociologists – was solidified.*

This concept percolated through the popular imagination. One day, in the late 1990s, graduate student Duncan Watts was asked by his father if he realised that he was only six handshakes away from the president. Watts, working for mathematician Steven Strogatz at the time, brought up this idea as they were discussing how groups of crickets could communicate. After this chance conversation, 'small world' would go from a quaint

---

* Milgram's experiment has been criticised by later researchers for lacking rigour. He didn't, for example, take into account the folders that never made it to the target person. As a result, whether six really is the magic number when it comes to degrees of separation between people has remained an open question. Luckily, data from social media sites is offering new ways to answer it.

expression to a mathematically defined property of a network.

In 1998, Watts and Strogatz published a paper laying out just what it takes for a graph to function like a small world. The notion of a short average path length – the idea that any two nodes are separated by only a few steps – was a key component of it. One way to get short path lengths is to make the graph heavily interconnected, *i.e.* a graph where each node connects directly to many others. This trick, however, is in clear violation of what we know about social networks: the average citizen of America – a country of around 200 million at the time – had only about 500 acquaintances, according to Milgram.

So, Watts and Strogatz limited their network simulations to those with sparse connections, but varied exactly what those connections looked like. They noticed that it was possible to have short path lengths in a network that was highly *clustered*. A cluster refers to a subset of nodes that are heavily interconnected, like the members of a family. In these networks, most nodes form edges just with other nodes in their cluster, but occasionally a connection is sent to a node in a distant cluster. The same way a train between two cities makes interactions between their citizens easier, these connections between different clusters in a network keeps the average path length low.

Once these characteristics were identified in their models, Watts and Strogatz went looking for them in real data – and found them. The power grid system of the United States – turned into a graph by considering any generator or substation as a node and transmission lines as edges – has the low path length and high clustering of a small world network. A graph made of actors with edges between any pairs that have starred in a movie together is the same. And the final place that they looked for, and found, a small world network was in the brain.

More specifically, the structure Watts and Strogatz analysed was the nervous system of the tiny roundworm, *Caenorhabditis elegans*. Ignoring the directionality of the neural connections, Watts and Strogatz treated any connection as an edge and each of the 282 neurons in the worm's wiring diagram as a node. They found that any two neurons could be connected by a path with, on average, only 2.65 neurons in between them and that the network contained far more clustering than would be expected if those 282 neurons were wired up randomly.

Why should the nervous system of a nematode have the same shape as the social network of humans? The biggest reason may be energy costs. Neurons are hungry. They require a lot of energy to stay in working order and adding more or longer axons and dendrites only ups the bill. A fully interconnected brain is, thus, a prohibitively expensive brain. Yet if connections become too sparse the very function of the brain – processing and routing information – breaks down. A balance must be struck between the cost of wiring and the benefit of information sharing. Small worlds do just this. In a small world, the more common connections are the relatively cheap ones between cells in a local cluster. The pricey connections between faraway neurons are rare, but there are enough to keep information flowing. Evolution, it seems, has found small worldness to be the smart solution.

Watts and Strogatz's finding in the roundworm was the first time the nervous system was described in the language of graph theory. Putting it into these terms made visible some of the constraints that are shared by the brain and other naturally occurring networks. Connections can be expensive to maintain, be they acquaintanceships or axons, and if these similarities exist between the roundworm and social networks it's reasonable to expect that the structure of other nervous systems is dictated by them as well.

But to speak about the structure of the nervous system requires that we know something about the structure of the nervous system. As it turns out, harvesting this information is a nuisance at best and an unprecedented technical hurdle at worst.

★ ★ ★

A 'connectome' is a graph describing the connections in a brain. While Watts and Strogatz were working with an incomplete version, the full roundworm connectome is defined by the full set of 302 neurons in the worm and the 7,286 connections between them. The roundworm was the first animal to have its entire adult connectome documented and, at present, it is the only one.

This lack of connectomic data is due largely to the gruelling process by which it is collected. Mapping a full connectome at the neuron level requires fixing a brain in a preservative, cutting it into sheets thinner than a strand of hair, photographing each of these sheets with a microscope and feeding those photographs into a computer that recreates them as a 3D stack. Scientists then spend tens of thousands of hours staring at these photographs, tracing individual neurons through image after image and noting where they make contact with each other.* The process of unearthing the delicate structure of neural connections this way is as painstaking as a palaeontology dig. The price of all this slicing, stitching and tracing makes it unlikely that full connectomes are within reach for any but the smallest of species. The connectome of the fruit fly, an animal with a brain one-millionth the size of a human's, is currently being assembled by a team of scientists and producing millions of gigabytes of data in the process. And, while any two roundworms are more or less alike, more complex species tend to have more individual differences, making the connectome of only a single fly or mammal a mere draw from a hat of possible connectomes.

Luckily more indirect methods are available that allow for a rough draft of connectomes in many individuals and species. One approach involves recording from a neuron while electrically stimulating others around it. If stimulation of one of these nearby neurons reliably causes a spike in the recorded one, there's likely a connection between them. Another option is tracers: chemicals that act like dyes that colour in a neuron. To see where inputs are coming from or outputs are going to, one just needs to look at where the dye shows up. None of these methods can create a complete connectome, but they do work to give a snapshot of connectivity in a certain area.

---

* There are currently some successful attempts to automate this arduous process. In the meantime, desperate scientists have also tried to turn this work into a game and get people all over the world to play it. It can be found at eyewire.org.

While connectivity had been studied long before it, the word 'connectome' wasn't coined until 2005. In a visionary paper, psychologist Olaf Sporns and colleagues called on their fellow scientists to help build the connectome of the human brain, promising it would 'significantly increase our understanding of how functional brain states emerge from their underlying structural substrate'. Getting connection data from humans is a staggering challenge, as many of the invasive methods used in animals are, for obvious reasons, not permissible. A quirk of brain biology, however, allows for a clever alternative.

When the brain builds connections, protecting the cargo is key. Like water leaking out of a seeping hose, the electrical signal carried by an axon is at risk of fading away. This isn't much of a problem for short axons connecting nearby cells, but those carrying signals from one region of the brain to another need protection. Long-range axons therefore get wrapped in layers and layers of a waxy blanket. This waxy substance, called myelin, contains a lot of water molecules. Magnetic resonance imaging (the same technology used to take pictures of tumours, aneurysms and head injuries) can detect the movement of these water molecules – information that is used to reconstruct the tracts of axons in the brain. Through this, it's possible to see which brain regions are connected to each other. After the publication of Sporns's plea, the Human Connectome Project was launched to map out the brain using this technique.

Identifying long-range axons this way doesn't produce the same kind of connectome that single-cell tracing methods do. It requires that scientists chunk the brain into rough and possibly arbitrary regions; it's therefore a much coarser description of connectivity. In addition, measuring water molecules isn't a perfect way to pick up the axons between these areas, leading to errors or ambiguities. Even David Van Essen, one of the key scientists of the Human Connectome Project, warned the neuroscience community in 2016 that there are major technical limitations to this approach that shouldn't be underappreciated. On the other hand, it is one of the only means by which we can peer into a living human brain; the desire to press forward with it makes sense. As Van Essen wrote: 'Be optimistic, yet critical of glasses half full and half empty.'

Despite these limitations, neuroscientists in the early 2000s were inspired by the work of Watts and Strogatz to look at their field through the lens of graph theory and eagerly set their gaze upon any and all available connectome data. What they saw when they analysed it were small worlds in all directions. The reticular formation, for example, is an ancient part of the brain responsible for many aspects of bodily control. When a cell-level map of this region in cats was pieced together and analysed in 2006, it was the first vertebrate neural circuit to be given the graph-theory treatment. And it was found to be a small world. In studies of the connections between brain areas in both rats and monkeys, short path lengths and plenty of clusters were always found as well. Humans were finally brought into the small-world club in 2007 when researchers in Switzerland used MRI scans to parcellate the brain into a thousand different areas – each about the height and width of a hazelnut – and measured the connections between them.

Universal findings are a rare sight in neuroscience; the principles that operate on one set of neurons aren't necessarily assumed to show up in another. As a conclusion that repeats itself across species and scales, small worldness is thus remarkable. Like the refrain of a siren song, it also implores more exploration. To see small worlds in so many places raises questions about how they got there and what roles they can play. While the answers to these questions are still being explored, without the language of graph theory they couldn't have even been asked in the first place.

★ ★ ★

On 10 February 2010, approximately 23 per cent of all flights originating in the United States were cancelled. This historically large disruption was the result of a snowstorm in the north-east that closed a handful of airports, including Ronald Reagan in Washington DC and JFK in New York. Such a sizable dent in travel wouldn't ordinarily come from closing a handful of airports – but these were not just any airports, they were hubs in the aviation network.

Hubs are nodes in a graph that have a high degree – that is, they're highly connected. They reside in the tails of the degree distribution: a plot that shows, for each degree value, how many nodes in the network have that degree (see Figure 21). For graphs like the aviation network or the structure of servers that make up the internet, this graph starts high – meaning that there are many nodes that have only a small number of connections – and fades as the number of connections increases, leading to a long,

Degree distribution

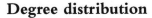

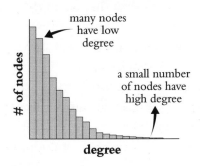

degree

An example hub

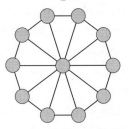

The node at the centre has
degree 10 while the rest
have degree 3

Figure 21

low tail that represents the small number of nodes
with very high degree, such as JFK airport. The high
degree of hubs makes them powerful but also a potential
vulnerability. Like removing the keystone from a stone
archway, a targeted attack on one of its hubs can cause a
network to collapse.

The brain has hubs. In humans, they're found sprinkled
throughout the lobes. The cingulate, for example, which
curves around the centre of the brain, serves as a hub; as
does the precuneus, which sits atop the back of the
cingulate.* In studies of sleep, anaesthesia and people in
comas, activity in these areas correlates with consciousness.
The size of another hub, the superior frontal cortex,
appears correlated with impulsivity and attention. Lesions
of a fourth hub, located in the parietal cortex on the side

---

* A note of caution: precisely what features a brain area needs to
have to be considered a hub is debated. And even applying the same
definition to different datasets can lead to different conclusions. As
it is still early days for this style of analysis, such kinks will take time
to work out.

of the brain, cause patients to lose a sense of direction. In total, the population of hubs appears diverse in both location and function. The connective thread, if there is any between them, is just how complicated each one is. Regions of the brain like the visual cortex, auditory cortex and olfactory bulb – regions with clear and identifiable roles present right there in their names – don't make it on to the list of hubs. Hub regions are complex; they pull in information from multiple sources and spread it out just as far. Their role as integrators seems a clear result of their placement in the network architecture.

Hubs, in addition to integrating information in distinctive ways, may also be responsible for setting the brain's clock. In the CA3 region (the memory warehouse of the hippocampus mentioned in Chapter 4), waves of electrical activity sweep through the neural population in the early days of development after birth. These waves ensure that the activity of the neurons and the strength of their connections are established correctly. And hub neurons are the likely coordinators of this important synchronised activity; they tend to start firing before these waves and stimulating them can instigate a wave. Other studies have even posited a role for hub regions in synchronising activity across the whole brain. Because of their high degree, a message from a hub is heard far and wide in the network. Furthermore, hubs in the brain tend to be strongly interconnected with each other, a network feature referred to as a 'rich club'. Such connections can ensure all the hubs are themselves on the same page as they send out their synchronising signals.

Even the way hub neurons develop points to their special place in the brain. The neurons that go on to form the rich club in the roundworm, for example, are some of the first to appear as the nervous system grows. In just eight hours after an egg is fertilised, all of these hub neurons will be born; the rest of the nervous system won't be finished until over a day later. Similarly, in humans, much of the basic hub structure is present in infants.

If hubs are so central to brain function, what role might they play in *dys*function? Danielle Bassett explored this question as part of her far-ranging career at the intersection of networks and neuroscience.

In the early 2000s, when the methods of graph theory were first being unfurled across the field of neuroscience, Bassett was a college student studying physics. At that time, it might have been a surprise to her to hear she'd go on to be called the 'doyenne of network science' by a well-known neuroscientist.[*] Though working towards a physics degree was itself already somewhat surprising given her upbringing: Bassett was one of 11 children home-schooled in a religious family where women were expected to play more traditional roles. Her transition to neuroscience came during her PhD, when she worked with Edward Bullmore, a neuropsychiatrist at Cambridge University who was part of an initial wave of neuroscientists eager to apply graph theory to the brain. One of Bassett's first projects was to see how the structure of the brain is affected by the common and crippling mental disorder schizophrenia.

Schizophrenia is a disease characterised by delusions and disordered thought. Comparing the brains of

---

[*] British neuroscientist Karl Friston bestowed this title on Bassett in a 2019 interview in *Science*. We'll hear more about Friston in Chapter 12.

people with the disease to those without, Bassett found several differences in their network properties, including in the hubs. Regions in the frontal cortex that form hubs in healthy people, for example, don't do so in schizophrenics. A disruption to the frontal cortex and its ability to rein in and control other parts of the brain could relate to the hallucinations and paranoia schizophrenia can elicit. And while the brain of a schizophrenic is still a small world, both the average path length and the strength of clustering are higher than in healthy people, making it seemingly more difficult for two disparate areas to communicate and get on the same page.

As the first study to approach this disease from the perspective of graph theory, this work helped bring an old idea about 'disconnection syndromes' into the quantitative age. As early as the late nineteenth century, neurologists hypothesised that a disruption in anatomical connections could lead to disorders of thought. German physician Carl Wernicke in particular believed that higher cognitive functions did not reside in any single brain region, but rather emerged from interactions between them. As he wrote in 1885: 'Any higher psychic process … rested on the mutual interaction of … fundamental psychic elements mediated by means of their manifold connections via the association fibres.' Lesioning these 'association fibres', he posited, would impair complex functions like language, awareness and planning.

Now that the tools of graph theory have met the study of 'disconnection syndromes', more diseases of this kind are being explored with modern approaches. One common example is Alzheimer's disease. When the brain-wide connectivity of older patients with Alzheimer's was compared to those without it, it was found that those with Alzheimer's had longer path lengths between brain areas. The confusion and cognitive impairment of Alzheimer's disease may result, in part, from the breakdown of efficient communication between distant brain regions. Similar changes in brain network structures are seen to a lesser extent with normal ageing.

Since starting her own lab at the University of Pennsylvania in 2013, Bassett has moved on from just observing the structure of the brain in health and disease to figuring out how to exploit it. The activity of complex networks can be hard to predict. A rumour whispered to a friend could die out immediately or spread across your social network, depending on the structure of that network and your friend's place within it. The effects of stimulating or silencing neurons can be equally hard to anticipate. The Bassett lab is combining tools from engineering with knowledge about the structure of brain networks to make control of neural activity more tractable. In particular, models based on the brain-wide connectomes of individual people were used to determine exactly where brain stimulation should be applied in order to have the desired effect. The aim is to use this individualised treatment to get disorders like Parkinson's disease and epilepsy under control.

The hope that the metrics of graph theory may serve as markers of disease – potentially even early markers that could lead to preventative care – has made them quite popular in medical research. Thus far, the brain disorders scrutinised by network analysis include Alzheimer's, schizophrenia, traumatic brain injury, multiple sclerosis, epilepsy, autism and bipolar disorder. Results, however, have been mixed. As was pointed out, the MRI technique for collecting the data in the first place has its problems and some studies find disease signatures that others don't. Overall, with so many excited scientists on the hunt for differences between diseased and healthy brains, some false positives and faulty data are bound to get caught up in the findings. But whether the results are solidified yet or not, it's safe to say this new slate of tools has made an arrival on the clinical scene.

★ ★ ★

A developing brain is an eruption. Neurons bubble out of a neuronal nursery called the ventricular zone at a breakneck pace and pour into all corners of the burgeoning brain. Once there, they start making connections. These indiscriminate neurons form synapse after synapse with each other, frenetically linking cells near and far. At the height of synapse building in the human brain – during the third trimester of pregnancy – 40,000 such connections are constructed every second. Development is an explosion of neuronal and synaptic genesis.

But just as soon as they come, many of these cells and connections go. An adult has far fewer neurons than they had in the womb; as many as half the neurons produced during development die. The number of connections a neuron in the cortex makes peaks around the first year of life and gets reduced by a third thereafter. The brain is thus built via a surge and a retreat, a swelling and a shrinking. During development, the pruning of neurons and synapses is ruthless; only the useful survive. Synapses, for example, are built to carry signals between neurons. If no signal flows, the synapse must go. Out of this turmoil and turnover emerge operational neural circuits. It's like encouraging the overgrowth of shrubbery for the purpose of carving delicate topiary out of it.

Such is the way biology found to build the brain. But if you asked a graph theorist how to make a network, they'd give the exact opposite answer. The designer of a public transit system, for example, wouldn't start by building a bunch of train stations and bus stops and connecting them all up just to see what gets used. No government would approve such a waste of resources. Rather, most graphs are built from the bottom up. For example, one strategy graph theorists use is to first build a graph that – using the fewest edges possible – has a path between any two nodes. This means some paths may be quite long, but by observing what paths get used the most (by commuters on a train or information travelling between servers on the internet) the network designer can identify where it would be useful to add a shortcut. Thus, the network gets more efficient by adding well-placed edges.

The brain, however, doesn't have a designer. There is no central planner that can look down and say: 'It looks like signals would flow better if that neuron over there was connected to this one here.'* That is why the brain needs to overproduce and prune. The only way the brain can make decisions about which connections should exist is by calculating the activity that passes through those connections. Individual neurons and synapses have elaborate molecular machinery for measuring how much use they're getting and growing or shrivelling as a result. If the connection doesn't exist in the first place, though, the activity on it can't be measured.

Pruning of connections in the brain starts off very strong, with synapses getting slashed left and right, but it then slows down over time. In 2015, scientists at the Salk Institute and Carnegie Mellon University explored why this pattern of pruning may be beneficial to the brain. To do so, they simulated networks that started overgrown and were pruned via a 'use it or lose it' principle. Importantly, they varied just how quickly this pruning happened. They found that networks that mimicked the pruning process in the brain (with the rate of pruning high initially and lowering over time) ended up with short average path lengths and were capable of effectively routing information even if some nodes or edges were deleted. This efficiency and robustness wasn't as strong in networks where the pruning rate was constant, or increased, over time. A decreasing pruning rate, it seems, has the benefit of quickly eliminating useless connections while still

---

* There may be an exception to this in very simple animals, such as *C. elegans*, where it's believed a lot of the information about who should connect to whom is encoded in the genome and is thus 'designed' through eons of natural selection.

allowing the network enough time to fine-tune the structure that remains; a sculptor working with marble, similarly, may be quick to cut out the basic shape of a man, but carving the fine details of the body is a slow and careful process. While most physical networks like roads or telephone lines will never be built based on pruning, digital networks that don't have costs associated with constructing edges – such as those formed by the wireless communication between mobile devices – could benefit from brain-inspired algorithms.

★ ★ ★

Network neuroscience, the name given to this practice of using the equipment of graph theory and network science to interrogate the structures of the brain, is a young field. *Network Neuroscience*, the first academic journal dedicated solely to this endeavour, was first published in 2017. New tools for mapping out connectomes at multiple scales have coincided with the computational power to analyse larger and larger datasets. The result is an electrified environment, with ever more and diverse studies of structure conducted every day.

A reason for caution, however, may be found in the stomach of a lobster.

The stomatogastric ganglion is a circuit of 25–30 neurons located in the gut of lobsters and other crustaceans. These neurons conspire through their connections to perform a basic but crucial job: produce the rhythmic muscle contractions that guide digestion. Eve Marder, a professor at Brandeis University in

Massachusetts, has spent half a century studying this handful of neurons.

Marder was born and raised in New York but her education took her to Massachusetts and then California.* While her doctorate work at the University of California, San Diego, was solidly in neuroscience, Marder always had an aptitude for mathematics: in primary school, she worked her way through maths textbooks meant for students two years her senior. This polymath personality permeates her science. Throughout her career she has collaborated with researchers from many backgrounds, including Larry Abbott (mentioned in Chapter 1), as he was making his transition from a particle physicist to renowned theoretical neuroscientist. Blending experimental exactness with a mathematical mindset, Marder has thoroughly probed the functioning of this little lobster circuit both physically and in computer simulations.

The connectome of the lobster stomatogastric ganglion has been known since the 1980s. The 30 neurons of this ganglion form 195 connections and send outputs to the muscles of the stomach. For her PhD, Marder worked out what chemicals these neurons use to communicate. In addition to standard neurotransmitters – the chemicals that traverse the small synaptic cleft between the neuron releasing them and the one receiving

---

* Marder entered graduate school in 1969, a time at which women were becoming a more common sight in these programmes, but barriers still existed. As she recounts in her autobiography, 'I knew it unlikely that I would get into Stanford biology because they were widely said to have a quota on women (2 out of 12).'

them – Marder also found a panoply of neuromodulators at play.

Neuromodulators are chemicals that fiddle with the settings of a neural circuit. They can turn the strength of connections between neurons up or down and make neurons fire more, less or in different patterns. Neuromodulators effect these changes by latching on to receptors embedded in a neuron's cell membrane. Part of what's noteworthy about neuromodulators is where they come from and how they get to the neuron. In the most extreme case, a neuromodulator could be released from a different part of the brain or body and travel through the blood to its destination. Other times, a neuromodulator is released locally from nearby neurons – but whether from near or afar, neuromodulators tend to bathe a circuit indiscriminately, touching many neurons and synapses in a diffuse way. Whereas regular neurotransmission is like a letter sent between two neurons, neuromodulation is a leaflet sent out to the whole community.

In the 1990s, Marder, along with members of her lab and the lab of Michael Nusbaum, a professor at the University of Pennsylvania, experimented with neuromodulators in the stomatogastric ganglion circuit. Ordinarily, the circuit produces a steady rhythm, with certain neurons in the population firing about once per second. But, when the experimenters released neuromodulators on to the circuit, this behaviour changed. Some neuromodulators made the rhythm increase: the same neurons fired, but more frequently. Others made the rhythm decrease, and some had a more dramatic effect, both disrupting the rhythm and activating neurons that were ordinarily silent. The neuromodulators causing these changes were all released from neurons that normally provide input to this circuit. This means these different patterns of output are plausibly produced naturally throughout the animal's life. In more artificial settings, neuromodulators added by the experimenters can cause even larger and more diverse changes.

Importantly, throughout these experiments the underlying network never changed. No neurons were added or deleted, nor did they cut or grow any connections. The marked changes in behaviour stemmed solely from a small sprinkling of neuromodulators atop a steady structure.

The massive effort poured into getting a connectome presupposes a certain amount of payoff that will come from having it, but the payoff is less if the structure-function relationship is looser than it may have seemed. If neuromodulators can release the activity of neurons in a circuit from the strict constraints of their architecture, then structure is not destiny. Perhaps this wouldn't be such a concern if neuromodulation were a phenomenon specific to the stomatogastric ganglion. This, however, is far from the truth. Brains are constantly bathing in modulating molecules. Across species, neuromodulators are responsible for everything from sleeping to learning, moulting to eating. Neuromodulation is the rule, not the exception.

Through mathematical simulations of the circuits she studies, Marder has explored not just how different behaviours arise from the same structure, but also how *different* structures can produce the *same* behaviours. Specifically, each lobster has a slightly different configuration of its gut circuitry: connections may be built stronger or weaker in one animal versus another. By simulating as many as 20 million possible ganglion circuits, Marder's lab found that the vast majority aren't capable of producing the needed rhythms, but certain specific configurations are. Each lobster, through some combination of genes and development, finds its way to one of these functioning configurations. The work makes an important point about individual brains: diversity doesn't always mean difference. What may look like a deviation from the structural norm could in fact be a perfectly valid way to achieve the same outcomes. That these diverse structures create the same rhythms adds another wrinkle to the structure–function relationship.

As much as Marder's work shows the limits of structure for understanding function, it also shows the need for it. Her lifetime of work – and all the insights it has provided – is built atop the connectome. Without that detailed structural information, there is no structure-function relationship to be explored. As Marder wrote in 2012: 'Detailed anatomical data are invaluable. No circuit can be fully understood without a connectivity diagram.' However, she goes on to remark that 'a connectivity diagram is only a necessary beginning, but not in itself, an answer'. In other words, when it comes to understanding the brain, knowing the structure of the nervous system is both completely necessary and utterly insufficient.

So, it may not be possible to satisfy Cajal's vision of intuiting the function of the nervous system from mere meditations on its structure. But the work of finding and formalising that structure is still an important prerequisite for any further understanding of the brain. Innovative methods for gathering connectome data are blossoming and the formalisms of graph theory are in place, prepared to take in and digest that data.

CHAPTER TEN

Making Rational Decisions

Probability and Bayes' rule

When Hermann von Helmholtz was a small child in early nineteenth-century Prussia, he walked through his hometown of Potsdam with his mother. Passing a stand containing small dolls aligned in a row, he asked her to reach out and get one for him. His mother did not oblige, however, though not out of neglect or discipline. Rather, she couldn't reach for the dolls because there were none. What the young Helmholtz was experiencing was an illusion; the 'dolls' he saw near him were actually *people* far away, at the top of the town's church tower. 'The circumstances were impressed on my memory,' Helmholtz later wrote, 'because it was by this mistake that I learned to understand the law of foreshortening in perspective.'

Helmholtz went on to become a prominent physician, physiologist and physicist. One of his greatest contributions was the design of the ophthalmoscope, a tool to look inside the eye that is still used by doctors to this day. He also furthered understanding of colour vision with his work on the 'trichromatic theory' – the idea that three different cell types in the eye each respond to different wavelengths of light – through which he deduced that colour-blind patients must lack one of these cell types. Outside the eye, Helmholtz published a volume on acoustics, the experience of tones, how sound is conducted through the ear and the way it excites the nerves. By turning his characteristic thoughtfulness and precision to the study of the sense organs, Helmholtz repeatedly illuminated the physical mechanisms by which information from the world enters the mind.

But the deeper question of how the mind uses that information always lingered with him. Inheriting a keen interest in philosophy from his father, Helmholtz's worldview was impacted in several ways by the work of German philosopher Immanuel Kant. In Kant's philosophy, the 'Ding an sich', or 'thing-in-itself', refers to the real objects out in the world – objects that can't be experienced directly, but only through the impressions they make on our sense organs. But if two different conditions in the world – for example, a nearby doll or faraway person – can produce the same pattern of light hitting the eye, how does the mind decide which is the correct one to perceive? How, Helmholtz wanted to know, can perception form out of ambiguous or uncertain inputs?

Ruminating on this question, Helmholtz concluded that a large amount of processing must go on between the point at which sensory information comes in and the moment it becomes a conscious experience. The result of this processing, he wrote, is 'equivalent to a *conclusion*, to the extent that the observed action on our senses enables us to form an idea as to the possible cause of this action'. This idea became known as 'unconscious inference' because the objects in the world must be inferred by their effects on the sense organs. Taking further inspiration from Kant, Helmholtz proposed that this inference proceeded by interpreting the current sensory input in light of pre-existing knowledge about the world. In particular, just as his mistake with the dolls taught him about perspective, Helmholtz believed that experiences in the past can influence perceptions in the present.*

Despite being one of the most mathematically adept physiologists of all time, Helmholtz never defined unconscious inference mathematically. His ideas on the topic, while thorough, remained qualitative and mostly speculative. They were also rejected. Scientists at the time felt that the notion of 'unconscious inference' was a contradiction in terms. Inference, or decision-making, was a conscious process by default; it simply couldn't be occurring beneath the surface.

But Helmholtz would be vindicated, nearly 100 years after his death, by psychologists using mathematics originally developed more than 50 years before his birth. Unconscious inference − dressed in the equations of probability − would come to encapsulate the basic mechanisms of how humans perceive, decide and act.

★ ★ ★

It's not uncommon for mathematical topics − even some of the most abstract − to have their origins in very practical professions. The tools of geometry arose from building and land surveying; ancient astronomers contributed to the concept of zero becoming commonplace; and the field of probability was born out of gambling.

---

* In this way Helmholtz deviated from Kant, who believed that much of this world knowledge was innate rather than learned.

Girolamo Cardano was an Italian physician but, not unlike many educated men of the sixteenth century, he felt comfortable dabbling in a variety of subjects. According to his own count, he wrote well over a hundred books – most of them lost to time – with titles as far-ranging as *On the Seven Planets*, *On the Immortality of the Soul* and *On the Urine*. Regarding one of his books that would remain for posterity, Cardano wrote: 'The book *On Games of Chance* I also wrote; why should not a man who is a gambler, a dicer, and at the same time an author, write a book on gaming?' And a gambler Cardano was; the book reads more like a gaming manual borne from personal experience than a textbook. Yet it was, for its time, the most thorough treatment of the rules of probability available.

Cardano focuses most of his mathematics on the casting of dice. He is quick to acknowledge that any of the six sides of a die is as likely as the others to show up, but that in practice they won't always be found equally: 'In six casts each point should turn up once; but since some will be repeated, it follows that others will not turn up.' After working through examples of what to expect when rolling one, two or three dice, he concludes: 'There is one general rule, namely, that we should consider the whole circuit, and the number of those casts which represents in how many ways the favourable result can occur, and compare that number to the remainder of the circuit.' In other words, the probability that a certain result will happen can be calculated as the number of outcomes that lead to that result divided by the total number of possible outcomes.

Take, for example, the rolling of two dice. If the rolling of one die has six possible outcomes, the rolling of two dice has 6 x 6 = 36 possible outcomes. If we say that our desired result is that, after rolling the two dice, their faces sum to three, then there are two possible outcomes that lead to that result: 1) the first die shows one and the second shows two; or 2) the first die shows two and the second shows one. The probability that we get our desired result is thus 2/36 or 1/18.

According to Cardano: 'Gambling is nothing but fraud and number and luck.' So, in addition to discussing the numbers, he made sure to devote more than two chapters to the topic of fraud. Much of the focus was on how to notice a cheat: 'The die may be dishonest either because it has been rounded off, or because it is too narrow (a fault which is plainly visible).' The book also gave tips on how to handle a cheat when you spot one: 'When you suspect fraud, play for small stakes, have spectators.' Though it should be noted that Cardano's autobiography offers a rather different take on how to react. Under the chapter entitled 'Perils, Accidents and Persistent Treacheries' he recalls a time when he noticed a man's cards were marked and 'I impetuously slashed his face with my poniard, though not deeply'.

Importantly, Cardano made clear that most of his calculations of probabilities held only if the dice in use were honest, not if he were playing 'in the house of a professional cheat' (as he described the incident above). In that case, the probabilities would need to be 'made so much the larger or smaller in the proportion to the departure from true equality'.

Accounting for different probabilities under different conditions – such as a cheating player – would later become known as *conditional probability*. Conditional probability can be thought of as a simple if-then statement. If you're given the fact that X is true, then what's the chance that Y is too? For example, *given* that the die is fair, the probability that a roll of it will produce a two is 1/6. Alternatively, the probability would be, say, 1/3 *given* that you're playing with a cheat who has altered the die to prefer twos. The probability of an event thus depends on the circumstances it is *conditioned upon*.

One topic that flummoxed mathematicians for centuries after Cardano was the question of *inverse probability*. Standard probability may be able to say how different dice create different chances, but the goal of inverse probability was to go the other way – to reverse the reasoning and find the cause behind the effects.* For example, if Cardano didn't know if he was in a game with a cheat or not, he could observe the rolls of the die to try to determine if it was biased. If one too many twos came up, he may have suspected that something was amiss (though hopefully he would have kept his poniard to himself).

French mathematician Pierre-Simon Laplace worked on the issue of inverse probability intermittently over 40 years of his career. The culmination came in 1812 with the publication of *Théorie Analytique des Probabilités*. Here,

---

* Speaking about probability in terms of cause and effect was not uncommon at the time. But it is not, in general, a wise thing to do. The probability that you'll be holding an umbrella given that the person next to you on the street is may be high, but the one is not the cause of the other.

Laplace demonstrates a simple rule that would come to be one of the most important and influential findings in mathematics.

The rule says that if you want to know the probability that the die is weighted, you have to combine two different terms. The first is the probability that the rolls you've seen could come from a weighted die and the second is the probability of the die being weighted in the first place. More technically, this is usually stated as: the probability of your hypothesis ('the die is weighted') given your evidence (the rolls you've seen) is proportional to the probability of your evidence given your hypothesis (the odds you'd see those rolls if the die were weighted) times the probability of your hypothesis (how likely is the die to be weighted to begin with) (see Figure 22).

Let's say the die has come up 'two' three times in a row and you want to know if you're being taken for a ride. With a fair die, the probability of that streak is 1/6 x 1/6 x 1/6 = 1/216. This would be called the probability of the evidence *given* the belief that the die is fair. On the other hand, the die may be weighted to roll a two, say, one-third of the time. The probability of the evidence

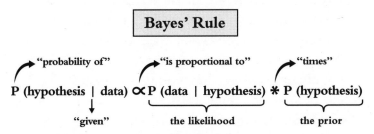

Bayes' Rule

"probability of"   "is proportional to"   "times"

P (hypothesis | data) ∝ P (data | hypothesis) * P (hypothesis)

"given"   the likelihood   the prior

The probability of the hypothesis given the data is proportional to the probability of the data given the hypothesis times the probability of the hypothesis

Figure 22

given the hypothesis the die is weighted this way would be 1/3 x 1/3 x 1/3 = 1/27. Comparing these numbers, it's clear that three twos in a row is much more probable with a weighted die than with a fair one; it seems you may be playing with a cheat.

But these numbers are insufficient. To draw a proper conclusion, the rule says we need to combine this with more information. Specifically, we need to multiply these numbers by the probability, in general, of the die being weighted or not.

Let's say in this case, your gambling partner is your closest friend of many years. You'd put the chance that they are using a weighted die at only 1 in 100. Multiplying the probability of getting three twos when using a weighted die times the low probability that the die is weighted, we get 1/27 x 1/100 = 1/2,700 or 0.00037. Doing this for the other hypothesis – that the die is unweighted – we get 1/216 x 99/100 = 0.0045. The second number being larger than the first, you'd be fair in concluding that your friend is not, in fact, a fraud.

What this example shows is the power of the *prior*. The 'prior' is the name given to the probability of the hypothesis – in this case the probability your friend has altered the die. Running through the same equations, but assuming you are playing with a stranger who is as likely to cheat as not (that is, the probability of cheating is 0.5), the outcome is different: 0.019 versus 0.0023 in favour of a weighted die. In this way, a strong prior can be a deciding factor.

The other term – the probability of the rolls given the hypothesis – is called the 'likelihood'. It indicates how likely you'd be to see what you've seen if your hypothesis about the world were true. Its role in inverse probability reflects the fact that, to determine the cause of any effect, one must first know the likely effects of each cause.

Both the likelihood and the prior on their own are incomplete. They represent different sources of knowledge: the evidence you have here and now versus an understanding accumulated over time. When they agree, the outcome is easy. Otherwise, they exert their influence in proportion to their certainty. In the absence of clear prior knowledge, the likelihood dominates the decision. When the pull from the prior is strong, it can make you hardly believe your own eyes. In the presence of a strong prior, extraordinary claims can only be believed with extraordinary evidence.

'When you hear hoof beats, think of horses, not zebras' is a bit of advice frequently given to medical students. It's meant to remind them that, of two diseases with similar symptoms, the more common one should be their first guess. It is also an excellent example of the rule of inverse probability in action. Whether you're in the presence of a horse or of a zebra, you've got a similar chance of hearing hoof beats; in technical terms, the likelihoods in the two cases are the same. Given such ambiguous evidence, the decision falls into the hands of the prior and, in this case, prior knowledge says horses are more common and therefore the best guess.

In the 200 years since the publication of his work, in papers, in textbooks and on classroom chalkboards, the equation for inverse probability that Laplace wrote down has been referred to as 'Bayes' rule'. Thomas Bayes was a Presbyterian minister in eighteenth-century England. Also an amateur mathematician, Bayes did work on the problem of inverse probability and he was able to solve a specific version of it. But all his thoughtfulness and calculations never quite got him to the form of Bayes' rule we know today. What's more, Bayes himself never published the work. An essay containing his thoughts on 'a problem in the doctrine of chances' was eventually sent to the Royal Society by a friend of his, another minister named Richard Price, in 1763, two years after Bayes' death. Price put substantial work into turning Bayes' notes into a proper essay; he wrote an introduction motivating the problem and added an extensive technical appendix (unfortunately all this effort did not prevent the essay from being referred to as 'one of the more difficult works to read in the history of statistics'*). Despite Laplace being alive at the time Bayes' essay was published, he did not seem to be aware of it until after he had made substantial progress on his own.

It could thus be said that the Reverend Bayes doesn't quite deserve the empire that's been posthumously gifted to him. It's not clear he would've wanted it anyway. Bayes' rule has not always fared well among scientists and philosophers. Much like Helmholtz's work on unconscious inference, the equation has variably been underutilised and misunderstood. This was, initially, due to the difficulty of applying it. Laplace himself was able to use the rule on some problems of measurement in astronomy and also to

---

* The statistician and historian responsible for this assessment, Stephen Stigler, is also known for 'Stigler's law of eponymy', which claims that a scientific law is never named after its true originator. The sociologist Robert Merton is believed to be the originator of this law.

support the long-running hypothesis that slightly more male than female babies are born on average. But, depending on the problem in question, applying Bayes' rule could involve some complex calculus, making it an onerous approach before the modern computer was around to help.

But the real struggle for Bayes' rule came later – and ran deeper. While the validity of Laplace's equation was unquestioned, how to *interpret* that equation occupied and divided statisticians for decades. According to philosopher of science Donald Gillies: 'The dispute between the Bayesians and the anti-Bayesians has been one of the major intellectual controversies of the twentieth century.' The biggest target in the crosshairs of the anti-Bayesians was the prior. Where – they wanted to know – does this information come from? In theory, it is world knowledge. In practice, it is someone's knowledge. As the giant of twentieth-century statistics Ronald Fisher said, the assumptions that go into choosing the prior are 'entirely arbitrary, nor has any method been put forward by which such assumptions can be made even with consistent uniqueness'. Without providing an unbiased, repeatable procedure for reaching a conclusion, Bayes' rule was no rule at all. Because of this, the method was cast aside, branded – in a way that would be sure to scare off serious scientists – as 'subjective'.

Conceptual concerns have a habit of fading when exposed to the light of practical proof, however. And in the latter part of the twentieth century, Bayes' rule was proving its worth. Actuaries, for example, came to realise that their rates were better calculated using the principles of inverse probability. In epidemiology, Bayes helped sort out the connection between smoking and lung cancer. And in the fight against the Nazis, code-breaker Alan Turing turned to Bayesian principles to uncover messages written with the 'unbreakable' Enigma machine. Bayes' rule was emerging as a tamer of uncertainty anywhere that it reared its head. Practically speaking, the prior proved only a minor problem. It could be initialised with an educated guess and updated in light of new evidence (or, barring any knowledge at all, each hypothesis is simply given an equal chance). With its repeated success in spite of an active movement against it, Bayes' rule certainly earns the title bestowed on it by Sharon McGrayne's book, *The Theory That Would Not Die*.

★ ★ ★

When Bayes' rule entered psychology it was not with a bang, but with a battering. No single publication carried it in. Rather, starting with the field of decision theory in the 1960s, multiple different lines of research employed it and explored it, until eventually the idea of the brain as Bayesian bloomed at the turn of the twenty-first century.

Some of the early work on Bayesian principles in the brain came from an unlikely place: space. On its mission to get space travel off the ground, the National Aeronautics and Space Administration (NASA) knew that it would have to engineer more than just flight suits and jet engines. It also investigated the 'human factors' of flying – such as how pilots read flight equipment, sense their environment and interact with controls. Researching this problem in 1972, aeronautical engineer

Renwick Curry wrote one of the earliest papers placing human perception in Bayesian terms. Specifically, he used Bayes' rule to explain patterns in how humans perceive motion. Academic boundaries being what they are, however, few psychologists heard about it.

Economics offered another way for Bayes to creep in. Ever eager to capture human behaviour in compact mathematical form, economists turned to Bayes' rule as early as the 1980s. In 'Are individuals Bayesian decision-makers?', written by William Viscusi in 1985, workers were shown to either over-or underestimate the riskiness of specific jobs because they relied on their prior beliefs about how risky jobs are in general.

Psychologists also spotted Bayes on the horizon via one of their former sources of inspiration. As we saw in Chapter 3, the study of the brain has been influenced by the field of formal logic. By the end of the twentieth century, in many ways, probability was the new logic – an improved way to assess how humans think. Instead of the harsh true-false dichotomy of Boolean logic, probability offers shades of grey. In this way, it aligns better with our own intuitions about our beliefs. As Laplace himself wrote: 'Probability theory is nothing but common sense reduced to calculation.'

Of course, probability is a bit better than that because the rules are mathematically worked out to be the *best* form of common sense – and Bayes' rule in particular is a prescription for how best to reason.

It was on these grounds that John Anderson formally debuted a Bayesian approach to psychology, under a method he referred to as 'rational analysis'. It was an idea that came to him in 1987 while he was in Australia, on sabbatical from his job as a professor of psychology and computer science at Carnegie Mellon University. Rational analysis, according to Anderson, stems from the belief that 'there is a reason for the way the mind is'. Specifically, it posits that an understanding of how the mind works will best grow out of an understanding of where it came from. When it comes to Bayes, the reasoning starts with the fact that humans live in a messy, uncertain world. Yet – Anderson argues – humans have evolved within this world to behave as rationally as possible. Bayes' rule is a description of how to reason rationally under conditions of uncertainty. Therefore, humans should be using Bayes' rule. Put simply, if evolution has done its job, we should see Bayes' rule in the brain.

The details of just how the rule will be applied and to what problems depend on more specific features of the environment. As an example, Anderson offers a Bayesian theory of memory recall. It says that the probability of a particular memory being useful in a particular situation is found by combining: 1) how likely you'd be to find yourself in that situation if that memory were useful; with 2) a prior that assumes more recent memories are more likely to be useful. This choice of prior is meant to reflect the fact that humans come from a world where information has a shelf life; therefore, more recent memories are more likely to be of value.

Importantly, under the rational analysis framework, 'rational' can be far from perfect. Memory, for example, certainly can fail us. But, according to this viewpoint, if we forget a fact from primary school 20 years on, we are not being irrational. Given the limited capacity of memory and the ever-changing world in which we live, it makes perfect sense to let old and little-used information go. In this way, the prior in a Bayesian model can be thought of as storing a shortcut. It's an encoding of the basic stats of the world that can make decision-making faster, easier and – in most cases – more accurate. If, however, we find ourselves in a world that deviates from the one we've evolved and developed in, our priors can be misleading. 'Think of horses' is only good advice in a place with more horses than zebras.

★ ★ ★

In early 1993, a group of researchers met at the Chatham Bars Inn in Chatham, Massachusetts. The group included psychologists David Knill (a professor at the University of Pennsylvania who served as organiser) and Whitman Richards (a professor at MIT who was part of the first crop of PhD students in the Department of Psychology there in the 1960s). Also at the meeting were scientists trained in physiology and neuroscience like Heinrich Bülthoff, whose work was on the visual system of fruit flies; as well as engineers and mathematicians such as Alan Yuille, a student of Stephen Hawking.

On the agenda for this eclectic bunch was a hunt for a new formal theory of perception – ideally one that could capture the complexities of the senses while also offering new, testable hypotheses. A complexity of particular concern was how the senses appear to be affected by more than just what meets the eye, ear or nose. That is, incoming sensory information combines with a rich set of background knowledge before perception is complete. According to Knill, no theory at the time was able to say precisely 'how prior knowledge should be brought to bear upon the interpretation of sensory data'.*

The meeting birthed a book, published in 1996, the title of which reveals the solution the attendees settled on: *Perception as Bayesian Inference.* The seeds for this idea had, as we've seen, been scattered around for some time, growing in different ways in different fields. This was an opportunity to pull them together. The book presents a unified and clear approach to the Bayesian study of perception, focusing mainly on the sense of vision. Its success spawned countless research papers in the years that followed. If Anderson's work on 'rational analysis' put Bayes on psychology's map, this book gave it its own country.

To understand the basics of Bayesian perception, consider an example. Light reflects off a flower and hits the eye. The wavelength of the light is around 670 nanometres (nm). It's the task of the brain to figure out, given the wavelength it's receiving, what the 'thing-in-itself' is, or what is really going on in the world. In Bayesian terms, this would be the probability of the hypothesis that a certain flower is present given that a wavelength of 670nm is hitting the eye.

Bayes' rule tells us what to do. First, we need to find out how likely we are to see that wavelength under different conditions. The likelihood of seeing 670nm light if the flower is blue and illuminated by white light

---

* Some of the existing theories of the time that failed to do this include the models of the visual system discussed in Chapter 6.

is very low (blue light falls between 450 and 480nm). The likelihood of seeing 670nm light if the flower is red and illuminated by white light is quite high; 670nm is right in the middle of the red spectrum. However, the probability of seeing 670nm light if the flower is *white* and illuminated by *red* light is also quite high. Since both of these scenarios are just as likely to produce 670nm light, if we stop here, we may be quite unsure of which one is the better interpretation.

But as good Bayesians, we remember the importance of the prior. The probability of a world illuminated by red light is, by most measures, quite small. White light, however, is a very common sight. The scenarios above that assume white light are thus much more probable. Multiplying the prior probability of different scenarios times the probability of seeing 670nm light in that scenario, we see that only one scores high on both of these measures. We therefore conclude that in front of us there is a red flower, illuminated by regular white light.

Of course, 'we' don't actually conclude that. This process, just as Helmholtz anticipated, happens unconsciously. The odds are weighed out of our sight and we know only the end result. It is, in this way, a never-ending procedure to produce perception – an underground production line in the mind. At each moment, probabilities are calculated and compared, each perception a bit of computation according to Bayes' rule.

With all the work that goes into perception, it's no surprise that the brain can sometimes come up with odd results. In 2002, a team of researchers out of the US and Israel catalogued a series of common illusions that people fall prey to when trying to estimate the movement of an object. It included the fact that the shape of an object influences the direction we think it is moving in, that two items moving in different directions may appear as one and that dimmer objects appear to move more slowly.

This may seem simply like a list of our failings, but the researchers found that all of these lapses could be explained by a simple Bayesian model. Particularly, these habits fall out of the calculation if we assume a specific prior: that motion is more likely to be slow than fast. Take, for example, the last illusion. When an object is hard to see, the evidence it provides about its movement is weak. In the absence of evidence, Bayes' rule relies on the prior – and the prior says things move slowly. This bit of mathematics may explain why drivers have a tendency to speed up in the fog – with weak information about their own movement, they assume their speed is too slow. Importantly, the Bayesian approach recasts these tricks of the mind as traits of a rational calculation. It shows how some mistakes are actually reasonable guesses in an uncertain world.

There is, however, another part to the process of perception. So far, we've simply assumed that the percept we experience should be the one with the highest probability. That's a reasonable choice, but it is a choice nonetheless, and a different one could be made.

Consider the Necker cube. This famous illusion (see Figure 23) admits more than one interpretation: the lower square could be seen as coming forwards, like a box faced slightly downwards, or it could be behind the plane of the page, suggesting a box tilted upwards. Both boxes are equally likely to produce this pattern of lines, so the

Necker Cube

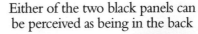

Either of the two black panels can
be perceived as being in the back

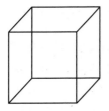 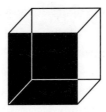 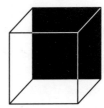

Figure 23

decision about which is the true state would be strongly
influenced by the prior. Let's say, downwards-tilted boxes
are a bit more likely in general. So, after applying Bayes'
rule, the probability of there being a downwards box
when we see these lines is 0.51 and an upwards box 0.49.
Taking the standard approach to mapping this to
perception, we'd say the larger of the two probabilities
wins – we see it as a downwards box, end of story.

On the other hand, rather than choosing one
interpretation and sticking to it, the brain could choose
to alternate between the two. The box could appear
downwards at one moment and upwards the next,
switching back and forth repeatedly. In this case, the
probabilities tell us not which interpretation to stick
with, but rather *the amount of time* to spend in each.

This switching is exactly what researchers at the
University of Rochester (including David Knill) saw in
an experiment in 2011. The experimenters overlaid two
visual patterns such that it was unclear whether the first
was on top of the second or vice versa – that is, the
image could be interpreted two different ways. Asking
people to indicate when their perception of the image

switched from one view to the other, they could determine the amount of time spent seeing each. Assuming that either pattern was equally likely to be on top (that is, the prior probabilities are the same), Bayes' rule says people should spend 50 per cent of their time seeing it each way. And this is exactly what they found. But to really test the predictive powers of Bayes' rule, the scientists needed to move away from the 50-50 scenario. To do this, they manipulated the image to make one pattern appear slightly more on top than the other. This altered the likelihood – that is, the probability of seeing this image given that one or the other pattern truly was on top. The more they altered the image in this direction, the more time viewers spent seeing the preferred pattern on top – exactly in accordance with Bayes' rule.

As this study shows, a set of probabilities can be mapped to a perception in interesting ways – a mapping known by scientists as the decision function. Bayes' rule itself doesn't tell us what decision function to use; it only provides the probabilities. Perception could be collapsed to the interpretation with the highest probability, or it could not. Perception could be a sampling from different interpretations over time in accordance with their probabilities, or it could not. Overall, perception could be the result of any complex combination of probabilities. The output of Bayes' rule therefore provides a rich representation of sensory information, one that the brain can use in any way that seems most reasonable. In this way, probabilities mean possibilities.

Another benefit of thinking of the mind as dealing in probabilities is that it opens the door to quantifying a potentially elusive concept: confidence. Confidence is intuitively tied to evidence and certainty. When walking around in a dark room, where visual evidence is weak, we move slowly because we aren't confident we won't bump into a wall or table. In a brightly lit room, however, the strong influence of the clear visual evidence removes such doubts. The Bayesian confidence hypothesis formalises this intuition by saying that how confident a person is in their interpretation of the world is directly related to the probability of that interpretation given the evidence – that is, the output of Bayes' rule. In the dark room where evidence is low, so is the probability of any given interpretation of the room and, therefore, so is confidence.

Researchers from the UK tested just how well this Bayesian hypothesis matches the data in 2015. To do so, they asked people to look for a particular pattern in two different images that were quickly flashed one after the other. The subjects then reported which of the two images had the pattern and, importantly, how confident they were in their decision. The decisions and confidence of the humans were compared to the predictions of the Bayesian model and to predictions from two simpler mathematical models. Bayes' rule provided the better match for the majority of the data, supporting the Bayesian confidence hypothesis.

★ ★ ★

'In the laboratory, we like to simplify the enormous task of understanding how the brain works,' said Dora Angelaki in an interview in 2014. 'Traditionally, neuroscience has studied one sensory system at a time. But in the real world, this is not what happens.'

Angelaki, originally from Crete, is a professor of Neuroscience at New York University. She credits her background in electrical engineering for her desire to seek out the underlying principles of how things work. As part of her research, she is correcting neuroscience's bias towards simplicity by studying how the senses interact.

The particular senses Angelaki seeks to combine are visual and vestibular. The vestibular system provides the little-known sixth sense of balance. Tucked deep in the ear, it is composed of a set of tiny tubes and stone-filled sacs. Through the sloshing of fluids in the tubes and the movement of the stones, the system offers – like the liquid in a level – a measurement of the head's tilt and acceleration. This system works in tandem with the visual system to provide an overall sense of place, orientation and movement. When these two systems are out of whack, unpleasant sensations like motion sickness can occur.

In her effort to understand the vestibular system, Angelaki has borrowed methods from an uncommon source: pilot training. Subjects in her experiments are strapped into a seat on a moving platform like the kind used in flight simulators. The platform can give them brief bursts of acceleration in different directions. At the same time a screen in front of them gives a visual sense of movement in the form of dots of light flowing past them – a visual not unlike that of jumping to lightspeed in *Star Wars*. While pilot training generally keeps the physical and visual motion aligned, Angelaki uses this set-up to see what the brain does when they disagree.

Bayes' rule offers a guess about that. Treating the visual and vestibular inputs as separate streams of evidence about the same external world, the mathematics of probability provides a simple means for how to combine them. Instead of a single likelihood term – like in the standard Bayes' rule – the two likelihoods (one from each sense) are multiplied together. Let's say your task is to determine if you're moving to the left or to the right. To calculate the probability that you're actually moving to the right – given some vestibular and visual inputs – you'd multiply the likelihood that you'd see that visual input if you were moving to the right *times* the likelihood that you'd receive that vestibular input if you were moving to the right. To complete the process, this value then goes on to be multiplied by the prior probability of rightwards movement. The same can be done for leftwards movement, and the two are compared.

Just as a rumour transforms into a fact as you hear it from many different people, in Bayes' rule getting the same information from multiple senses strengthens the belief in that information. When the moving platform and the display screen are both consistent with rightwards movement, both visual and vestibular likelihoods will be high and thus so will the outcome of their multiplication. This contributes to a confident conclusion of rightwards movement. If they're put at odds – the platform moves right while the dots say left – then the vestibular likelihood would still say the probability of rightwards movement is high, but the visual one would say it's low. Multiplying these leads to a middling result and only moderate confidence one way or the other.

But, as with rumours, the reliability of the source matters. In her experiments, Angelaki can reduce the confidence subjects have in one or the other sensory inputs. To make the visual inputs less reliable, she simply makes them messier. That is, rather than all the dots moving together to give a strong sense of directed motion, some dots move randomly. The more dots that are random, the less reliable the visual information becomes.

Looking at how that plays out with probabilities, we see that Bayes' rule naturally titrates how much to rely on a source based on how reliable it is. If the dots were moving completely randomly, the visual input would provide no information about movement direction. In this case, the likelihood of the visual input given rightwards movement would be equal to the likelihood of it given leftwards movement. With equal likelihoods on both sides, the visual input wouldn't weigh the decision either way. The tiebreaker would come from the vestibular inputs (and the prior). If, instead, 90 per cent of the dots moved randomly and 10 per cent indicated rightwards movement, then the likelihood of the visual input in support of rightwards movement would be slightly higher than that of leftwards. Now the visual input does weigh the decision – but only slightly. As the visual input becomes more reliable, its say in the decision grows. In this way, Bayes' rule automatically puts more stock in a source in proportion to how certain it is.

Investigating the conclusions people come to about their movement in these experiments, Angelaki and her lab have shown once again that, for the most part, humans behave as good Bayesians. When the visual evidence is weak, they depend more on their vestibular system. One caveat is that, while they use the visual information more as it becomes more reliable, they still don't use it quite as much as Bayes' rule would predict. That is, the vestibular input is consistently overemphasised – an effect found in monkeys as well. This could be a result of the fact the visual input is always a bit ambiguous: seeing dots moving past could be an effect of your own movement, or it could just be the dots moving. So, the vestibular input is, in general, a more trustworthy source and therefore worthy of more weight.

★ ★ ★

Once the Bayesian approach to perception was unleashed, it quickly spread to all corners of psychology. Like a Magic Eye illusion, staring at any data long enough can make the structure of Bayes' rule pop out of it. As a result, priors and likelihoods abound in the study of the mind.

As we've already seen, Bayes' rule has been evoked to explain motion perception, the switching of ambiguous illusions like the Necker cube, confidence and the combination of vision and vestibular inputs. It's also been adapted to account for how we can be tricked by ventriloquists, our sense of the passing of time and our ability to spot anomalies. It can even be stretched and expanded to cover such tasks as motor skill learning, language understanding and our ability to generalise. Such a unifying framework to describe so much of mental activity seems an unmitigated success. Indeed, according to philosopher of mind Michael Rescorla, the Bayesian approach is 'our best current science of perception'.

Yet, not all psychologists can be counted as devoted disciples of the Reverend Bayes.

To some, a theory that explains everything is at risk of explaining nothing at all. The flip side of the flexibility of the Bayesian approach is that it can also be accused of having too many *free parameters*. The free parameters of a model are all its movable parts – all the choices that the researcher can make when using it. The same way that, if given enough strokes even the worst golfer could eventually get the ball in the cup, if given enough free parameters any model can fit any data. If a finding from a new experiment conflicts with an old one, for example, an over-parameterised model can easily twist its way around to encompass them both. If getting the model to fit the data is as easy as a bank making change, its success isn't very interesting. A model that can say anything can never be wrong. As psychologists Jeffrey Bowers and Colin Davis wrote in their 2012 critique of the Bayesian approach: 'This ability to describe the data accurately comes at the cost of falsifiability.'

There are indeed many ways to squeeze parts of perception into a Bayesian package. Take, for example, the likelihood calculation. Computing a quantity like 'the likelihood of seeing 670nm light given the presence of a red flower' requires some knowledge and assumptions about how light reflects off different materials and how the eye absorbs it. Without a perfect understanding of the physical world, the model-maker must put in some of their own assumptions here. They could, therefore, wiggle these assumptions around a bit to match the data. Another source of choice is the decision function. As we saw earlier, the output of Bayes' rule can be mapped to the perception and decision of an animal any number of ways. This option, too, has the power to make any action look Bayesian in theory. And then, of course, there are those pesky priors.

Just as they gave pause to statisticians in the twentieth century, priors have proven a challenge to psychologists of the twenty-first. If the assumption of a certain prior – for example, that motion is likely to be slow – helps explain a psychological phenomena, that can be taken as good evidence that the brain really uses that prior. But what if a different phenomena is best explained by a different prior – say, one that assumes motion is fast? Should it be assumed that the priors in our mind are constant across time and task? Or are they flexible and fluid? And how can we know?

As a result of these concerns, some researchers have embarked on an exploration into the properties of priors. French cognitive scientist Pascal Mamassian has worked to investigate a particularly common one: the assumption that light comes from above. For more than two centuries, experiments and illusions have found that humans keep this implicit belief about the source of illumination in mind as they make sense of shadows in a scene. It's a sensible guess, given the location of our dominant light source, the sun. More recently, experiments have revised this finding slightly and found that humans actually assume that light comes from above *and slightly to the left*. Mamassian conducted tests revealing this bias in the lab, but he also found a more creative way to interrogate it. Analysing 659 paintings from the Louvre museum in Paris, he found that in a full 84 per cent of portraits and 67 per cent of non-portrait paintings, the light source was indeed biased towards the left side. Artists may have come to prefer this setting exactly because it aligns with our intuitions, creating a more pleasant and interpretable painting.

Another open question about priors is their origin. Priors can serve as an efficient way to imprint facts about the world on our minds; but are these facts gifted to us from previous generations through our genes, or do we develop them in our own lifetimes? To test this, a study in 1970 raised chickens in an environment wherein all light came from below. If the prior assumption that light comes from overhead is learned in their lifetimes, these birds wouldn't have it. How the animals interacted with visual stimuli, however, showed they did still assume light should be from above. This points in favour of an inherited prior.

Humans, of course, aren't chickens, and the development of our nervous system may allow for a bit more flexibility. Investigating the prior biases of children of various ages, psychologist James Stone found in 2010 that children as young as four did show a bias towards assuming overhead light, though it was weaker than that in adults. This bias grows steadily over the years to reach adult strength, suggesting that a partially innate prior may be fine-tuned by experience. Further in support of this flexibility, a team from the UK and Germany showed in 2004 that our grip on where light must be coming from can be loosened. Through training, participants were able to shift their prior beliefs about the source of the light by several degrees.

Picking a particular prior and poking at it from different directions through a multitude of experiments helps verify it as a resilient and reliable effect. Each prior this is done for becomes less of a free parameter in the model and more fixed.

Another question supporters of the Bayesian brain hypothesis need to address is 'how?'

While there are reasons to believe the brain *should* use Bayes' rule, and there is evidence that it *does*, the question of how this plays out in neurons remains a lively area of research.

When it comes to priors, scientists are looking for what cupboard in the brain stores these bits of background knowledge and how they get mixed into the neural decision-making process. One hypothesis is that it's a simple numbers game. If a group of neurons are charged with representing something about the world – say, where in the environment a sound is coming from – each neuron may have its own preferred location. This means it responds the most when sound is coming from there. If the brain determines where the sound is by adding up the activity of all the neurons that prefer the same location, locations with more neurons will have an advantage. So, if the prior says that sound is more likely to come from central locations than the periphery, this can be implemented by simply increasing the number of neurons that prefer the centre. As it turns out, neuroscientists Brian Fischer and Jose Luis Peña found this exact scheme in the brains of owls in 2011. Identifying neural signatures of priors this way can give insights to where they come from and how they work.

Theorists are building – and experimentalists are testing – many more hypotheses about how Bayes' rule plays out in the brain. There are a multitude of ways that neurons can conspire to combine likelihoods and priors.

These different hypotheses should not be considered in competition, nor should any single winner expect to be crowned at the end. Rather, while Bayes' rule can be one-size-fits-all for capturing the outputs of perception, the physical underpinnings of this rule may come in many different shapes and styles.

CHAPTER ELEVEN

How Rewards Guide Actions

Temporal difference and reinforcement learning

For much of his life as a scientist, Ivan Petrovich Pavlov had one passion: digestion. He started his academic work in 1870 with a thesis on pancreatic nerves. For 10 years as a professor of pharmacology in St Petersburg, he devised ways of measuring gastric juices in animals as they went about their life to show how secretions from different organs change in response to food or starvation. And by 1904, he was granted the Nobel prize 'in recognition of his work on the physiology of digestion through which knowledge on vital aspects of the subject has been transformed and enlarged'.

It is surprising, then, given all of his success in studying the gut, that Pavlov would go down in history as one of the most influential figures in psychology.

Pavlov's transition to studying the mind was, in a way, accidental. In an experiment designed to measure how dogs salivate in response to different foods, he noticed their mouths watering before the food even arrived – all it took was the sound of the footsteps of the assistant bringing in the bowls. This was not completely unusual. Much of Pavlov's previous work looked at how the digestive system is influenced by the nervous system, but these were usually more obvious interactions such as the smell of food impacting stomach secretions – interactions that were plausibly thought to be innate to the animal. Drooling at the sound of footsteps isn't a response hardwired into genes. It has to be learned.

Pavlov was a strict and unforgiving scientist. When public shootings related to the Russian Revolution caused a colleague to be late to a meeting, Pavlov replied: 'What difference does a revolution make when you have experiments to do in the laboratory?' But this intensity could lend itself to meticulous work, and when he decided to follow up on these salivation observations he did so thoroughly and exhaustively.

Pavlov would repeatedly present a dog with a neutral cue – like the ticking of a metronome or the sound of a buzzer (but not a bell, as is commonly thought; Pavlov relied only on stimuli that could be precisely controlled). He followed this neutral cue with food. After these pairings he would observe how much the dogs salivated in response to the cue alone. He wrote, in characteristic detail: 'When the sounds from a beating metronome are allowed to fall upon the ear, a salivary secretion begins after nine seconds, and in the course of 45 seconds 11 drops have been secreted.'

Varying the specifics of this procedure, Pavlov catalogued many features of the learning process. He asked questions like: 'How many pairings of cue then food does it take to reliably learn?' (around 20); 'Does the timing between the cue and the food matter?' (yes, the cue has to start before the food arrives but not too much before); 'Does the cue need to be neutral?' (no, the animals could learn to salivate in response to slightly negative cues, such as the application of a skin irritant); and many more.

This process – repeatedly pairing an upcoming reward with something usually unrelated to it until the two become linked – is known as classical or (unsurprisingly) 'Pavlovian' conditioning and it became a staple in early psychology research. Reviewers of Pavlov's 1927 book outlining his methodology and results described his work as 'of vital interest to all who study the mind and the brain' and 'remarkable both from the point of view of the exactness of his methods and the scientific insight shown in the sweeping character of his conclusions'.

Pavlov's work eventually fed into one of the biggest movements in twentieth-century science: behaviourism. According to behaviourism, psychology should not be defined as the study of the mind, but rather, as the study of behaviour. Behaviourists therefore prefer descriptions of observable external activity to any theorising about internal mental activity like thoughts, beliefs or emotions. To them, the behaviour of humans and animals can be understood as an elaborate set of reflexes – that is, mappings between inputs from the world to outputs produced by the animal. Conditioning experiments like Pavlov's offered a clean way of quantifying these inputs and outputs, feeding into the behaviourism frenzy.

After the publication of his book, therefore, many scientists were eager to replicate and build off Pavlov's work. American psychologist B. F. Skinner, for example, learned about Pavlov through a book review written by famed sci-fi author H. G. Wells. Reading this article piqued Skinner's interest in psychology and set him on the path to becoming a leading figure of the behaviourist movement, conducting countless precise examinations of behaviour in rats, pigeons and humans.*

When any field of science amasses enough quantitative data, it eventually turns to mathematical modelling to make sense of it. Models find structure in piles of numbers; they can stitch together disparate findings and show how they arise from a unified process. In the decades after Pavlov, the amount of data being generated from behavioural experiments on learning made it ready for modelling. As William Estes, a prominent American psychologist working on the mathematics of learning, wrote in 1950, conditioning data 'are sufficiently orderly and reproducible to support exact quantitative predictions of behaviour'.

Another paper, published in 1951, agreed: 'Among the branches of psychology, few are as rich as learning in quantity and variety of available data necessary for model building.' This paper, 'A mathematical model for simple learning', was written by Robert Bush and Frederick Mosteller at the Laboratory of Social Relations at Harvard University. Bush was a physicist-turned-psychologist and Mosteller a statistician. Together, influenced by the work of Estes, they laid out a formula for learning associations between cues and rewards that would be the starting point for a series of increasingly elaborate models. Through the decades, the learning that these models capture became known as 'reinforcement

---

* The kind of conditioning Skinner is most associated with is known as 'operant conditioning', which involves performing an action before getting a reward. The line between operant and Pavlovian conditioning is sometimes sharp, sometimes blurred and information in this chapter will at times relate to both.

learning'. Reinforcement learning is an explanation for how complex behaviour arises when simple rewards and punishments are the only learning signals. It is, in many ways, the art of learning what to do without being told.

★ ★ ★

In their model, Bush and Mosteller focused on a specific measure of the learned association between the cue and reward: the probability of response. For Pavlov's dogs, this is the probability of salivating in response to the buzzer. Bush and Mosteller used a simple equation to explain how that probability changes each time the reward is – or isn't – given after the cue.

Say you start with any random dog off the street (it is, in fact, rumoured that Pavlov got his subjects by stealing them off the streets). The probability that this dog will salivate at the sound of a buzzer starts at zero; it has no reason to suspect that the buzzer means food. Now you press the buzzer and then give the dog a piece of meat. According to the Bush-Mosteller model, after this encounter, the probability that the dog will salivate in response to the buzzer increases (see Figure 24). The exact amount that it increases depends on a parameter in the formula called the learning rate. Learning rates control the speed of the whole process. If the learning rate is very high, a single pairing could be enough to solidify the buzzer-food relationship in the dog's mind. At more reasonable rates, however, the probability of salivating remains low after the first pairing – maybe it goes to 10 per cent – and raises each time the buzzer is followed by food.

Regardless of the value of the learning rate, however, the second time the buzzer is followed by food, the probability of salivating increases *less* than it did after the first time. So, if it went from 0 to 10 per cent after the first pairing, it would increase only another nine percentage points, to 19 per cent, after the second pairing. And only by about eight percentage points after the third. This reflects that, in the Bush-Mosteller model (and in the dogs), the change to the probability with each pairing depends on the value of the probability itself. In other words, learning depends on what is already learned.

This is, from a certain angle, intuitive. Nothing new is learned from seeing the sun rise every day. To the extent that we believe something will happen, its actual happening has little effect on us. Anticipated rewards are no different. We don't update our opinion of our boss, for example, if we get the same holiday bonus we've received for the past five years. And the dogs only update their response to the buzzer to the extent that the food that follows differs from what they expect. The power to change expectations comes only from violating them.

Acquiring a Cue-Reward Association	**Extinction of a Cue-Reward Association**

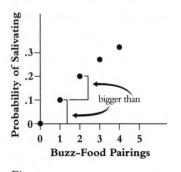 

Figure 24

This violation can be for better or worse. To the dog, the first piece of post-'buzz' meat is a lovely surprise and one that has a big impact on its expectations. After repeated pairings, though, expectations shift and slobbering at the sound of the buzzer becomes second nature. At this point, the most impactful thing that could happen would be to hear the buzzer and *not* receive food. Such a deprivation would lead to a large decrease in the probability of future salivation – a decrease as large as the increase that occurred with the first pairing. This inverse side of reward-based learning, wherein the animal learns to *dis*associate a cue from a reward, is called extinction. With each presentation of the cue without the expected reward, the extinction process unbuilds the association, eventually extinguishing the learned response entirely. Bush and Mosteller made a point of showing that their model captures this process to a tee as well.

At the same time that Bush and Mosteller were turning salivation information into equations, another man on the opposite side of the country was working to apply mathematics to some of the trickiest problems in business and industry. The deep and important connections between these works wouldn't be realised for decades to come.

* * *

The RAND Corporation is an American think tank founded in 1948. A non-profit offshoot of the Douglas Aircraft Company, its central aim was to extend the collaboration between science and the military that blossomed out of necessity during the Second World War. The corporation's name is appropriately generic (RAND literally stands for Research ANd Development) for the range of research projects it pursues. Over the years, RAND employees have made significant contributions to the fields of space exploration, economics, computing and even foreign relations.

Richard Bellman worked at RAND as a research mathematician from 1952 to 1965. An admirer of the subject as early as his teenage years, Bellman's path to becoming a mathematician was repeatedly interrupted by the Second World War. First, to lend support to the war effort, he left his postgraduate training at Johns Hopkins University to teach military electronics at the University of Wisconsin. He later moved to Princeton University, where he taught in the Army Specialized Training Program and worked on his own studies. He would eventually complete his PhD at Princeton, but not before he was drafted to work in Los Alamos as a theoretical physicist for the Manhattan Project. The intrusions didn't seem to impact his career prospects much. He became a tenured professor at Stanford University just three years after the war at the age of only 28.

Leaving the academic world for RAND at 32 was, in Bellman's words, the difference between being 'a traditional intellectual, or a modern intellectual using the results of my research for the problems of contemporary society'. At RAND his mathematical skill was applied to real-world problems. Problems like scheduling medical patients, organising production lines, devising long-term investment strategies or determining the purchasing plan for department stores. Bellman didn't have to set foot in a hospital or on a factory floor to help with these issues, however. All of these problems – and a great many more – are huddled under one abstract mathematical umbrella.

And, in the eyes of a mathematician, to be able to solve any of them is to solve them all.

What these problems have in common is they are all 'sequential decision processes'. In a sequential decision process, there is something to be maximised: patients seen, items produced, money made, orders shipped. And there are different steps that can be taken to do that. The goal is to determine which set of steps should be taken. How can the maximum be reached? What is the best way to climb the mountain?

Without much previous work to draw on in this field, Bellman turned to a tried-and-true strategy in mathematics: he formalised an intuition.[*] The mathematical conclusion this led him to is now known as the Bellman equation and the simple intuition it captures is that the best plan of action is the one in which all the steps are the best possible ones to take. Obvious though it may seem, when written in mathematics even banal statements can have power.

To see how Bellman made use of this intuition, we have to understand how he framed the problem. Bellman first set out to define how good a plan was in terms of how much reward – be that money, widgets, shipments, *etc.* – it is likely to accrue. Let's say you have a five-step plan. The total reward is the sum of the reward you get at each of those five steps. But after you've taken the first step, you now have a four-step plan. So, we could instead say that the total reward from the original five-step plan is the reward you get from taking the first step plus the

---

[*] Interestingly, Bellman was aware of Bush and Mosteller's publications, but his work on these problems was developed independently of that.

total reward from the four-step plan. And the total reward from the four-step plan is just the reward from taking *its* first step plus the reward from the resulting three-step plan. And so on, and so on.

By defining the reward of one plan in terms of the reward of another, Bellman made his definition *recursive*. A recursive process is one that contains itself. Consider, for example, alphabetisation. If you want to alphabetise a list of names, you'd start by sorting all the names according to their first letter. After that, you'd need to apply that *same* sorting process again on all the names starting with the same letter to sort them according to their second letter, and so on. This makes alphabetisation recursive.

Recursion is a common trick in mathematics and computer science in part because recursive definitions are flexible; they can be made as long or as short as needed. The formula for calculating a plan's total reward, for example, can just as easily be applied to a five-step plan as a 500-step one. Recursion is also a conceptually simple way to accomplish something potentially difficult. Like the turns of a spiral staircase, each step in a recursive definition is familiar but not identical, and we need only follow them down one by one to the end.

Bellman's framing contains two further insights that helped make his strategy effective for deploying on real-world problems. The first was to incorporate the very relatable fact that a reward you get immediately is worth more than a reward you get later. He did this by introducing a *discounting factor* into his recursive definition. So, whereas in the initial formula the reward from a five-step plan was equal to the reward from the first step plus the full reward from the four-step plan, an equation with discounting would say it is equal to the reward from the first step plus, maybe, 80 per cent of the reward from the four-step plan. Discounting is a way to weigh immediate gratification against delayed; it is 'a bird in the hand is worth two in the bush' codified into mathematics.

The second insight was more conceptual and more radical. It was a switch from focusing on rewards to focusing on *values*.

To understand this switch, let's consider the owner of a small business – a very small business. Angela is a busker on the New York City subway system. She knows she can play her electric violin for 20 minutes at certain subway stations before being chased away by the authorities, at which point she's not allowed to return. Different stations, however, have different payouts. Tourist areas can be very lucrative whereas commuter stops for native New Yorkers yield far fewer donations. She's leaving her house on Greenpoint Avenue in Brooklyn and wants to end up near a friend's place in Bleecker Street. What path should she take to make the most money on the way to her destination?

So far, we've noticed that, after starting from one position and taking a step in a plan, we find ourselves in circumstances broadly similar to how we began – except we are starting from a different position and have a different plan. In sequential decision-making, the different positions we can move through are called states and the steps in a plan are frequently referred to as actions. In Angela's case, the states are the different subway stations she can be at. Each time Angela takes an action (for example, from station A to station B), she finds herself in a new state (station B) that both yields some reward (the amount of donations her playing gets) and provides her with a new set of possible actions (other stations to go to). In this way, states define what actions are available (you can't go straight from Greenpoint Avenue to Times Square, for example) and actions determine what the next states are.

This interplay – wherein the actions taken as part of a plan affect what actions will be available in the future – is part of what makes sequential decision processes so difficult. What Bellman did was to take this constellation of states, actions and rewards, and turn it on its head. Rather than talk about the reward expected from a series of actions, he focused on the value that any given state has.

Value, as used colloquially, is a nebulous concept. It elicits ideas about money and worth, but also deeper notions of meaning and utility that can be hard to pin down. The Bellman equation, however, defines value precisely. Using the same recursive structure introduced earlier, Bellman defined the value of a state as the reward you get in that state plus the discounted value of the next state. You'll notice, in this definition there is no explicit concept of a plan; value is defined by other values.

Yet, this equation does rely on knowledge of the next state. Without a plan to say what action is taken, how do we know what the next state will be? This is where the original intuition – the idea that the best plan is made up of the best actions – comes into play. To calculate the value at the next state, you simply assume that the best possible action is taken. And the best possible action is the one that leads to the state with the highest value! When wrapped up in the language of value, the plan itself fades away.

So how does this help Angela? Given a map of possible subway stations (see Figure 25) and the associated donations she expects to get from each, we can calculate a 'value function'. A value function is simply the value associated with each state (in this case, each station). We can calculate this by starting at the end and working backwards. Once Angela reaches Bleecker Street, she will go straight to her friend's house and not do any busking, so the reward she will get at her final destination is $0. Because there are no further states from this point, the value of Bleecker Street is also zero. Backing up from here, the values of Union Square and 34th Street can be calculated in terms of the reward expected there and the value of Bleecker Street. This process continues until the value for each station is calculated.

With these values in hand, Angela can now plan her journey. Starting out from Greenpoint Avenue, she can take the train to either Court Square or Metropolitan Avenue. Which should she choose? Looking just at the possible rewards from each, Metropolitan Avenue seems the better choice, as it offers $10 versus Court Square's $5. But looking at the value function, Court Square is the correct choice. This is because the value function cares about what states you can get yourself into in the future – and from Court Square Angela can go straight to the jackpot, Times Square. Angela could also go to Queen's Plaza from Court Square, but that isn't relevant here, because the value function assumes Angela is smart. It assumes that from Court Square she would go to

Angela's Map of the Subway

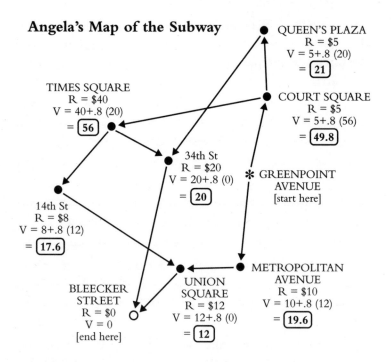

QUEEN'S PLAZA
R = $5
V = 5+.8 (20)
= $\boxed{21}$

TIMES SQUARE
R = $40
V = 40+.8 (20)
= $\boxed{56}$

COURT SQUARE
R = $5
V = 5+.8 (56)
= $\boxed{49.8}$

34th St
R = $20
V = 20+.8 (0)
= $\boxed{20}$

* GREENPOINT
AVENUE
[start here]

14th St
R = $8
V = 8+.8 (12)
= $\boxed{17.6}$

METROPOLITAN
AVENUE
R = $10
V = 10+.8 (12)
= $\boxed{19.6}$

BLEECKER
STREET
R = $0
V = 0
[end here]

UNION
SQUARE
R = $12
V = 12+.8 (0)
= $\boxed{12}$

'R' is the money earned at each station
'V' is the value of each station
With a discount of 80%: V = R + .8 V (of next station)

Figure 25

Times Square because Times Square is the better choice. All in all, following the value function would take Angela through Court Square to Times Square then to 34th Street and finally on to her destination at Bleecker Street. In total, she will have earned $65 – the most that any path on this map could offer.

Bellman's move to focus on the value function was important because it corrected a flaw in the original framing of the problem. We started out by trying to calculate the total reward we could get from a given plan. But when solving a sequential decision process, we aren't given a plan. In fact, a plan is exactly what we're trying to find! Once we know the value function, though, the plan is simple: follow it. Like breadcrumbs left on a forest path, the value function tells you where to go. Anyone looking for the most reward needs only to greedily seek the next state with the highest value. All actions can be chosen based on this simple rule.

Some interesting things happen as a result of the discounting that is part of the definition of value. For example, look at the options Angela has from Times Square. She can either go to 34th Street, get $20 and then end at Bleecker Street or she can go to 14th Street, get $8, then go to Union Square and get $12 and finally end at Bleecker Street. Both routes earn her $20 in total. But the value of 34th Street is 20 whereas the value of 14th Street is 17.6 (calculated as 8 + 0.8 x 12), indicating that 34th Street is the better option. This demonstrates how discounting future rewards can lead to plans with fewer steps; if there is only so much reward to get, it's best to get it quicker rather than slower. Discounting also means that even big rewards will be ignored if they are too distant. If a train station in New Jersey could garner Angela $75, it may still not influence her choice as she leaves her house. The impact of a reward on a value function is like the ripple from a stone dropped in water. It's felt strongest in the nearby states, but its power is diluted the farther away you go.*

---

* Because it controls the balance between caring about now versus the future, the strength of discounting can have sizable impacts on value and therefore on which actions are chosen. Scientists have posited that disorders such as addiction or ADHD can be understood in terms of inappropriate reward discounting. More on addiction later.

This technical definition of value – based on states and recursion and discounting factors – may seem a far cry from the word we use in everyday language. But those colloquial connotations are very much present in this equation. Why do we value money? Not because there is actually much pleasure to be had from the paper or the coin itself, but because of the future we can envision once we have that paper or coin. Money is worth only what it can get us later and what we can get later is baked into Bellman's definition of value.

Bellman's work to frame sequential decision processes this way truly did allow him to become the 'modern intellectual' he aimed to be when moving to RAND. In the years after his first publications describing this solution, countless companies and government entities began to apply it out in the world. By the 1970s, Bellman's ideas had been deployed on problems as diverse as sewer system design, airline scheduling and even the running of research departments at large companies such as Monsanto. The technique went under the name 'dynamic programming', a rather bland phrase Bellman actually coined with the aim of keeping some of the mathematics-phobic higher-ups in the military out of his hair. 'The 1950s were not good years for mathematical research,' Bellman wrote in his autobiography. 'The RAND Corporation was employed by the Air Force, and the Air Force had [Charles] Wilson as its boss, essentially. Hence, I felt I had to do something to shield Wilson and the Air Force from the fact that I was really doing mathematics inside the RAND Corporation. ... Thus, I thought dynamic programming was a good name. It was something not even a

Congressman could object to. So I used it as an umbrella for my activities.'

In applying the method in each of these settings, engineers had to find a way to calculate the value function. In some cases, like the subway example given above, the landscape of the problem is simple enough that the calculation is straightforward. But simple problems are rarely realistic. The real world has a large number of potential states; these states can connect to each other in complex or even uncertain ways; and they can do so through many possible actions. Much effort was put into finding the value function in these trickier situations. Yet even with clever techniques, the application of dynamic programming usually pushed at the edge of computing power at the time. The calculation of the value function was always a bottleneck in the process. And without a way to find the value function, the full potential of Bellman's contributions would remain unreached.

★ ★ ★

There is an irony to Pavlov's legacy. Its immediate effect was to set off behaviourism, a movement with religious-like dedication to ignoring the mind and focusing only on directly measurable behaviour. Yet the lineage of mathematical models it spawned found success in the other direction, by going increasingly inside the mind; to capture reinforcement learning in equations required the use of terms representing hidden mental concepts.

One of the well-known extensions to the Bush-Mosteller model came 20 years later, in 1972, and was developed by another duo, Yale psychologists Robert

Rescorla and Allan Wagner. Rescorla and Wagner generalised the Bush–Mosteller model, making it applicable to a wider range of experimental settings and able to capture more findings. The first alteration they made was to the very measure the model was trying to explain.

Bush and Mosteller's 'probability of response' was too specific and too limited. Rescorla and Wagner instead aimed to capture a more abstract value which they referred to as 'associative strength'. This strength of the association between a cue and a reward is something that would exist in the mind of the participant, making it not directly measurable, but different experiments could try to read it out in different ways. This could include measuring a probability of response, like the probability of salivating, but also other measures, such as the amount of salivation, or behaviours like barking or movement. In this way, Rescorla and Wagner subsumed the Bush–Mosteller model into a broader framework.

The Rescorla–Wagner model also expanded to incorporate a known feature of conditioning experiments referred to as 'blocking'. Blocking occurs when an initial cue is paired with a reward, and then a second cue is also given along with the first and both are paired with the reward. So, for example, after a dog had learned to associate the sound of the buzzer with food, the experimenter would then flash a light at the same time as the buzzer and then give the food. In Bush and Mosteller's model, the cues were treated completely separately. So, if the light and buzzer were paired with food enough times, the dog should come to associate the light with food the same way it had learned the association with the buzzer. Then we would expect that presenting the light alone would cause the dog to salivate. This is not, in fact, what happens; the dogs don't salivate in response to the light alone. The presence of the buzzer *blocks* the ability of the light to be associated with the food.

This provides further proof that learning is driven by errors. In particular, errors about predicted reward. When the animal hears the buzzer, it knows food is coming. So, when the food arrives there is no error in its prediction of that reward. As we saw before, that means it doesn't update its beliefs about the buzzer. But it also means it doesn't update its beliefs about anything else either. Whether there was a light on at the same time as the buzzer or not is irrelevant. The light has no bearing on the predicted reward, the received reward or the difference between the two, which defines the prediction error – and without an error, everything is stuck as it is. Prediction error is the grease that oils the wheels of learning.

Rescorla and Wagner thus made their update in associative strength between one cue and a reward dependent not just on that cue's current associative strength, but on the sum of the associative strengths of all cues present. If one of these associative strengths is high (for example if the buzzer is present), then the presence of the reward wouldn't change any of them (the association with the light is never learned). This summing across multiple cues is also something that would need to be done internal to the animal, further reflecting the rejection of behaviourism and a move to the mind.

But the watershed moment in reinforcement learning came in the mid-1980s from the work of a ponytail-wielding Canadian computer scientist named Richard

Sutton and his PhD adviser Andrew Barto. Sutton was educated in both psychology and computer science, and Barto spent a lot of his time reading the psychology literature. This proved a powerful combination as the work they did together pulled from and gave back to both fields.

Sutton's work removed the final tangible element of the model: the reward itself. Up until this point, the moment of learning centred around the time that a reward was given or denied. If you catch a whiff of the smoke from a blown-out candle and are then handed a piece of birthday cake, the association between the two strengthens. But a candle extinguished at the end of a religious ceremony likely doesn't come with cake and so the association weakens. In either case, though, the cake itself is the important variable. Its presence or absence is key. Anything can serve as a cue, but the reward must be a primal one – food, water, sex. But once we come to associate smoke with birthday cake, we may notice some other regularities. For example, the smoke is usually preceded by singing and the singing may be preceded by people putting on silly hats. None of these things are rewards in and of themselves (especially the singing, at most parties), but they form a chain that attaches each in some degree to the primary reward. Knowing this information can be useful: if we want cake, being on the lookout for silly hats may help.

Rescorla and Wagner had no way of allowing for this backing up of associations – no way, essentially, for a cue associated with a reward in one circumstance to play the role of a reward in another. But Sutton did. In Sutton's algorithm – known as 'temporal difference learning' – beliefs are updated in response to any violation of expectations. Walking through your office hallway to your desk, for example, expectations about reward may be pretty low. But when you hear your co-workers in the conference room start the first verse of 'Happy Birthday', a violation has occurred. Beliefs must be updated; you are now in a state where reward is on the horizon. This is where temporal difference learning occurs. You may choose to enter the conference room, finish the song, smell the candles and eat the cake. In doing those actions, no further violations will have occurred – and therefore, no further learning. It is thus not the receipt of the reward itself that causes any changes. The only learning that happened was in that hallway, many steps away from the reward.

What exactly is being learned here, though? What mental concept is it that got updated in the hallway? It's not the association of a cue with a reward – not directly at least. Instead it's more of a signal telling you the path to reward, should you follow the right steps in between.

This may sound familiar because what temporal difference learning helps you learn is a value function. At each moment in time, according to this framing, we have expectations – essentially a sense of how far we are from reward – that define the value of the state we are in. As time passes or we take actions in the world we may find ourselves in new states, which have their own associated values. If we've correctly anticipated the value of those new states, then all's well. But if the value of the current state is different from what we predicted it would be when we were in the state before, we've got an error. And errors induce learning. In particular, if the value of the current state is more or less than we expected it to be when we were in the previous state, we *change the value of the previous state*. That is, we take the surprise that occurred now and use it to change our belief about the past. That way, the next time we find ourselves in that previous state, we will better be able to predict the future.

Consider driving to an amusement park. Here the value of your location is measured in terms of how far you are from this rewarding destination. As you leave your house, you expect to arrive in 40 minutes. You drive straight for five minutes and get on a highway. You now expect to arrive in 35 minutes. After 15 minutes on the highway, you take an exit. Your estimated time of arrival is now 20 minutes. But, when you get off that exit and turn on to a side street you hit a traffic jam. As you sit in your hardly moving car, you know you won't be at the park for another 30 minutes. Your expected arrival time has now jumped up by 10 minutes – a significant error.

What should be learned from this error? If you had an accurate view of the world, you would've anticipated 30 more minutes of driving at the moment you took the exit. So, temporal difference learning says you should *update the value of the state associated with that exit*. That is, you use the information received at one state (a traffic jam on the side street) to update your beliefs about the value of the state before (the exit). And this may mean that the next time you drive to this amusement park, you will avoid that exit and choose another instead. But it doesn't take arriving at the amusement park 10 minutes late to learn from that mistake; the expectation of that happening at the sight of the traffic was enough.

What Sutton's algorithm shows is that by mere exploration – simple trial and error – humans, animals or even an artificial intelligence can eventually learn the correct value function for the states they're exploring. All it takes is updating expectations when expectations change – 'learning a guess from a guess' as Sutton describes it.

As an extension of Bellman's work on dynamic programming, temporal difference learning had the potential to solve real-world problems. Its simple moment-by-moment learning rule made it attractive from a computing perspective: it didn't demand as much memory as programmes that needed to store the entire set of actions that preceded a reward before learning from it. It also worked. One of the biggest displays of its power was TD-Gammon, a computer program trained via temporal difference learning to play the board game backgammon. Board games are particularly useful tests of reinforcement learning because rewards frequently only come at the very end of a game, in the form of a win or loss. Using such a coarse and distant signal to guide strategy at the very first move is a challenge, but one that temporal difference learning could meet. Built in 1992 by Gerald Tesauro, a scientist at IBM, TD-Gammon played hundreds of thousands of games against itself, eventually reaching the level of an intermediate player without ever taking instructions from a human. Because it learned in isolation, it also developed strategies not tried by humans (who are generally influenced by each other's gameplay to stick within a certain set of moves). In the end, TD-Gammon's unusual moves actually influenced theory and understanding about the game of backgammon itself.

In 2013 another application of temporal difference learning made headlines, this time applied to video games. Scientists at the artificial intelligence research company DeepMind built a computer program that taught itself to play multiple games from the 1970s arcade system Atari. This artificial gamer got the full Atari experience. The only inputs to the algorithm were the pixels on the screen – it was given no special knowledge that some of those pixels may represent spaceships or ping-pong bats or submarines. The actions it was allowed to take included the standard buttons such as up, down, left, right, A, B; and the reward for the model came in terms of the score provided by the game it was playing. As this burdens the algorithm with a more challenging task than backgammon – which at least had the concepts of pieces and locations baked into the inputs to the model – the researchers combined temporal difference learning with deep neural networks (a method we encountered in Chapter 3).[*] One version of this deep neural network had around 20,000 artificial neurons and, after weeks of learning, reached human-level performance on 29 out of 49 games tested. Because this Atari algorithm also learned asocially, it ended up with some interesting quirks, including discovering a clever trick for tunnelling through a wall in the brick-clearing game *Breakout*.

While games are a flashy and fun way to demonstrate the power of this approach, its application didn't stop there. After Google acquired DeepMind in 2014, it set reinforcement-learning algorithms to the task of

---

[*] Specifically, they used a deep convolutional neural network which, as we saw in Chapter 6, is used to model the visual system.

minimising energy use in its massive data centres. The result was a 40 per cent decrease in energy used to cool the centres and likely hundreds of millions in savings over the years as a result. With a single-minded focus on achieving the goal at hand, reinforcement learning algorithms find creative and efficient solutions to hard problems. These alien minds can thus help devise plans humans would've never thought of.

The paths of sequential decision-making and Pavlovian conditioning represent a victory of convergent scientific evolution. The trajectories of Bellman and Pavlov start with separate and substantial problems, each seething with their own demanding details. How should a hospital schedule its nurses and doctors to serve the most patients? What causes a dog to salivate when the sound of a buzzer hits its ears? These questions are seemingly worlds apart. But by peeling away the weight of the specifics – leaving only the bare bones of the problem to remain – their interlocking nature becomes clear. This is one of the roles of mathematics: to put questions disconnected in the physical world into the same conceptual space wherein their underlying similarities can shine through.

The story of reinforcement learning is thus one of successful interdisciplinary interaction. It shows that psychology and engineering and computer science can work together to make progress on hard problems. It demonstrates how mathematics can be used to understand, and replicate, the ability of animals and humans to learn from their surroundings. The story would be a remarkable one as it is, if it ended there. But it doesn't end there.

★ ★ ★

Octopamine is a molecule found in the nervous systems of many insects, molluscs and worms. It is so named because of its initial discovery in the salivary glands of the octopus in 1948. In the brain of a bee, octopamine is released upon run-ins with nectar. In the early 1990s, Terry Sejnowski, a professor at the Salk Institute in San Diego, California, and two of his lab members, Read Montague and Peter Dayan, were thinking about octopamine. In particular, they built a model – a computer simulation of bee behaviour – that was centred on the neuron in the bee brain that releases octopamine. A bee's choices about what flowers to land on or avoid, they proposed, could be explained by a Rescorla–Wagner model of learning and a neural circuit including the octopamine neuron could be the hardware that implements it. But as they worked out this octopamine puzzle, the team heard of another study, conducted some 6,000 miles away by a German professor named Wolfram Schultz, on octopamine's chemical cousin dopamine.

You may be familiar with dopamine. It has a bit of a reputation in popular culture. Countless news articles refer to it as something like 'our brain's pleasure and reward-related chemical' or talk about how everyday activities like eating a cupcake cause 'a surge of the reward chemical dopamine to hit the decision-making area of the brain'. It's branded as the pleasure molecule and it's not uncommon for products to be pedalled under its powerful name. Pop stars have named albums and songs after it. 'Dopamine diets' claim (without evidence) to provide foods that boost dopamine while keeping you slim. And the tech start-up Dopamine Labs promised to increase user engagement in phone apps by doling out squirts of the neurotransmitter. This poor celebrity chemical has also been badly maligned – referred to as the source of all addictions and maladaptive behaviours. Online communities like The Dopamine Project have cropped up aiming to provide 'better living through dopamine awareness'. And some Silicon Valley dwellers have even attempted 'dopamine fasts' as a respite from constant over-stimulation.

While it is true that a release of dopamine can accompany rewards, that is far from the whole story. What Schultz's study in particular showed was a case in which the neurons that dole out dopamine were *silent* when a reward was given.

Specifically, Schultz trained monkeys to reach their arm out in front of them in order to receive some juice.[*] During this training process, he recorded the activity of a population of dopamine-releasing neurons tucked into the underside of the brain. Schultz observed that at the end of training – when the animals knew they would get some juice by making a reach – these neurons showed no response at all to the juice reward being delivered.

When Schultz first published these results he didn't have a clear explanation for why the dopamine neurons were behaving this way, but the members of the Sejnowski lab did. And they reached out to Schultz to embark on a collaboration that would test the hypothesis that dopamine neurons encode the prediction errors necessary for temporal difference learning. It would be

---

[*] This is actually an example of the 'operant' form of conditioning mentioned earlier, because the animals need to make a reach to receive their reward.

the start of what Sejnowski referred to as 'one of the most exciting scientific periods of my life'.

Dayan and Montague worked to reanalyse Schultz's data through the lens of learning algorithms. They focused on the simplest of Schultz's experiments, which consisted of a light at the desired reach location turning on and, if the animal reached to it, a drop of juice was delivered half a second later. What they wanted to know was how the response of the dopamine neurons changed as the animal came to learn this association. But they were also interested in a particular circumstance after learning: what happens when the juice doesn't follow the light. If the animals learned the light–juice association, they would know to expect it and if the juice didn't show up that would be a significant prediction error. Did the dopamine neurons reflect that?

The neurons that release dopamine tend to fire around five spikes per second when nothing much is going on. At the start of the learning process, right after the animal got what seemed like a surprise shot of juice after making an arm movement, that rate jumped up briefly to about 20 spikes per second. The light that came before the movement, however, elicited nothing. But after enough pairings, once the animal came to understand how the light and the reach and the juice were all related, that pattern shifted. The dopamine neurons stopped responding to the juice. This is a change perfectly in line with the notion that they signal prediction error, because once the animal can correctly predict the juice there is no more error. And they started responding to the light. Why? Because the light had become associated with the reward but – crucially – they had no idea when it would come on. When it did arrive it was an error. Specifically, it's an error in the predicted value of the state of the animal. Sitting in the experimental chair going about its life, the monkey expects the next moment to be more or less similar to the current one. When the light turns on, that expectation is violated. Like hearing the first few bars of 'Happy Birthday' in your office hallway, it's a pleasant surprise, but a surprise nonetheless.

The final analysis – done while sporadically omitting the juice after the reach – was to see how *unpleasant* surprises were encoded. If dopamine is encoding errors, it should indicate when things are worse than expected as well. And with the juice absent, the neurons did just that. They had a dip in their firing right at the time the juice would've been delivered. Specifically, the neurons would go from five to 20 spikes per second in response to the light; then as the animal reached out its arm they'd return back to five. But, about half a second after the reach, when it was clear there was no juice coming, they'd shut off completely. An expectation had been violated and the dopamine neurons were letting it be known.

This study showed that the firing of dopamine neurons can signal the errors – both positive and negative – about predicted values that are needed for learning. It was thus an important point in shifting the understanding of dopamine from a pleasure molecule to a pedagogical one.

If the point of encoding the error is to learn from it, though, where does that learning happen? It turns out that's a bit hard to pin down because these dopamine-releasing neurons release dopamine in many corners of the brain; their projections burrow through the brain like pipework, touching regions near and far. Still, a location that seems particularly important is the striatum. The striatum is a group of neurons that serves as the primary input to a collection of brain areas involved in guiding movement and actions. Neurons in the striatum contribute to the production of behaviour by associating sensory inputs with actions or actions with other actions.

As we saw in Chapter 4, Hebbian learning is an easy way for associations between ideas to become encoded in the connections between neurons. Under Hebbian rules, if one neuron regularly fires before another, the weight of the connection from the first to the second is strengthened. In reinforcement learning, however, we need more than just to know that two events happened close in time. We need to know how those events relate to reward. Specifically, we only want to update the strength between a cue and an action (for example, seeing a light and reaching for it) if that pairing turns out to be associated with reward.

So, the neurons in the striatum don't follow basic Hebbian learning. Instead they follow a modified form wherein the firing of one neuron before another only strengthens their connection if it happens *in the presence of dopamine*. Dopamine – which encodes the error signal needed for updating values – is thus also required for the physical changes needed for updating that occur at the synapse. In this way, dopamine truly does act as a lubricant for learning.

Having the language of temporal difference learning in which to talk about the functioning of the brain has altered the conversation on clinical topics such as addiction. One theory, put forth in 2004 by neuroscientist

David Redish, tries to explain the addictive properties of drugs like amphetamine and cocaine in terms of the effects they have on dopamine release. It posits that these drugs cause a release of dopamine that is independent of the true prediction error. Specifically, by overdriving the dopamine neurons, these drugs send the false signal to the rest of the brain that the drug experience is always better than expected. This errant error signal still drives learning, pushing the estimated value of states associated with drug use higher and higher. Deforming the value function in this way is guaranteed to have detrimental effects on behaviour like the ones seen in addiction.[*]

★ ★ ★

David Marr was a British neuroscientist with a background in mathematics. His book, *Vision: A Computational Investigation into the Human Representation and Processing of Visual Information* was published in 1982, two years after his death. In the first chapter, he lays out the components needed for a successful analysis of a neural system. According to Marr, to understand any bit of the brain we should be able to explain it on each of three levels: computational, algorithmic and implementational. The computational level asks what is the overall purpose of this system, that is, what is it trying to do? The algorithmic level asks how, *i.e.*, through what steps, does it achieve this

---

[*] This theory can explain several aspects of addiction, but one of its big predictions failed. If these drugs lead to non-stop prediction error, then the blocking phenomena described previously shouldn't be seen when drugs are used as reward. An experiment in rats indicated, however, that blocking still does occur.

goal. And finally, the implementational level asks specifically what bits of the system – what neurons, neurotransmitters, *etc.* – carry out these steps.

An explanation encompassing all of Marr's levels is an aspiration towards which many neuroscientists strive. The systems that carry out reinforcement learning are a rare case where they can come within striking distance of this high bar. At the computational level reinforcement learning has a simple answer: maximise reward. This is what Bellman recognised as the goal of sequential decision processes and what following the value function should get you. But how do we learn the value function? That's where temporal difference learning comes in. The work of Bush, Mosteller, Resorla, Wagner and Sutton all turned stacks of data from conditioning experiments into strings of symbols that could describe the algorithm needed to do the learning part of reinforcement learning. On the implementation level, dopamine neurons take on the task of calculating prediction error and the signals they send to other brain areas control the associations learned there. In this way, a satisfying understanding of a fundamental ability – to learn from rewards – was achieved by tunnelling towards the topic from many different angles.

CHAPTER TWELVE

Grand Unified Theories of the Brain

Free energy principle, Thousand Brains Theory and integrated information theory

One of the biggest shockwaves in the history of science hit physics in the mid-nineteenth century. James Clerk Maxwell, a Scottish mathematician, published his seven-part paper 'A dynamical theory of the electromagnetic field' in 1865. Through this marathon of insightful analogies and equations, Maxwell demonstrated a deep and important relationship between two already important forms of physical interaction: electricity and magnetism. Specifically, by defining electromagnetic field theory, Maxwell created the mathematical infrastructure needed to see the equations of electricity and magnetism as two sides of the same coin. In the process, he concluded that a third important object – light – was a wave in this electromagnetic field.

Scientists had, of course, studied electricity, magnetism and light for centuries before Maxwell. And they had learned a fair bit about them, how they interact and how they can be harnessed. But Maxwell's unification provided something profoundly different – a whole new way to interpret the physical world. It was the first domino to fall in a series of landmark discoveries in foundational physics and paved the way for many of today's technologies. Einstein's work, for example, is built on electromagnetic field theory and he reportedly credited his success to standing 'on the shoulders of Maxwell'.

More than its direct impact on research, however, Maxwell's theory planted the thought in the minds of future physicists that more subterranean relationships may exist among physical forces. Digging up these connections became a major aim of theoretical physics. By the twentieth century an explicit search for what are known as grand unified theories (GUTs) was on. At the top of the to-do list was finding a GUT that could further unify electromagnetism with two other forces: the weak force (which governs radioactive decay) and the strong force (which holds atomic nuclei together). A big step in this direction came in the early 1970s with the discovery that the weak force and electromagnetism become one at very high temperatures. Even so, bringing in the strong and weak forces still leaves out another big one, gravity. Physicists, therefore, remain in pursuit of a full GUT.

GUTs tap into the aesthetic preferences of many physicists: simplicity, elegance, totality. They demonstrate how the whole can become greater than the sum of its parts. Before identifying a GUT, scientists are like the blind men touching an elephant in the old parable. They're each relying on what little information they can grab from the trunk, the leg or the tail. Through this they come up with separate and incomplete stories about what each bit is doing. Once the entire elephant is seen, however, the pieces lock into place and each is understood in context of the others. The deep wisdom obtained by finally finding a GUT can't be approximated by studying the parts separately. Therefore, as difficult as they can be to find, the search for GUTs is largely deemed a worthwhile endeavour by the physics community. As physicist Dimitri Nanopoulos wrote in 1979, shortly after helping to coin the phrase, 'grand unified theories give a very nice and plausible explanation of a whole lot of different and at first sight unrelated phenomena, and they definitely have the merit and right to be taken seriously'.

But should GUTs of the brain be taken seriously? The notion that a small number of simple principles or equations will be able to explain everything about the form and function of the brain is appealing for the same reasons GUTs are coveted in physics. However, most scientists who study the brain are doubtful they could exist. As psychologists Michael Anderson and Tony Chemero wrote: 'There is every reason to think that there can be no grand unified theory of brain function because there is every reason to think that an organ as complex as the brain functions according to diverse principles.' A GUT for the brain, as great as it would be, is considered a fantasy by many.

On the other hand, so much of what has been imported from physics to neuroscience – the models, the equations, the mindset – has helped advance the field in one direction or another. GUTs, core as they are to modern physics, are hard to ignore. They can be tantalising for those who study the brain even if they seem unlikely and to some scientists they are simply too seductive to pass up.

Searching for a GUT in the brain is a high risk–high reward endeavour. As such, it tends to require a big

personality to lead it. Most candidate GUTs of the brain
have a frontman of sorts – a scientist, usually the one
who first developed the theory, who functions as the
public face of it. To get a GUT to succeed also requires
dedication: supporters of a theory will work for years,
even decades, on refining it. And they're always on the
prowl for new ways of applying their theory to every
facet of the brain they can find. Advocacy is important,
too; even the grandest of GUTs can't help to explain
much if no one has heard of it. Therefore, many papers,
articles and books have been written to communicate
GUTs not just to the scientific community but to the
world at large. It's best for GUT enthusiasts to have a
thick skin, however. Pushing of such theories can be met
with disdain from the mass of scientists doing the more
reliable work of studying the brain one piece at a time.

Sociologist Murray S. Davis offered a reflection on
theories in his 1971 article entitled 'That's interesting!'
In it, he said: 'It has long been thought that a theorist is
considered great because his theories are true, but this is
false. A theorist is considered great, not because his
theories are true, but because they are *interesting* … In
fact, the truth of a theory has very little to do with its
impact, for a theory can continue to be found interesting
even though its truth is disputed – even refuted!' Grand
unified theories of the brain, whatever their truth may
be, are surely interesting.

★ ★ ★

Generally jovial and soft-spoken, British neuroscientist
Karl Friston doesn't quite fit the profile for the leader of

an ambitious and controversial scientific movement. Yet he certainly has the devoted following of one. Scientists – ranging from students to professors, including those well outside the traditional bounds of neuroscience – gather ritualistically on Mondays to receive a few moments of his insights each. They are there to seek his unique wisdom mainly on one topic. It is an all-encompassing framework on which Friston has been building an understanding of the brain, behaviour and beyond for over 15 years: the free energy principle.

'Free energy' is a mathematical term defined by differences between probability distributions. Yet its meaning in Friston's framework can be summarised somewhat simply: free energy is the difference between the brain's predictions about the world and the actual information it receives. The free energy principle says that everything the brain does can be understood as an attempt to minimise free energy – that is, to make the brain's predictions align as much as possible with reality.

Inspired by this way of understanding, many researchers have gone on a search for where predictions may be made in the brain and how they may get checked against reality. A small industry of research built around the idea of 'predictive coding' explores how this could be happening in sensory processing in particular.[*] In most predictive coding models, information gets sent as normal through the sensory processing system. For

---

[*] The predictive coding scheme was actually developed in the absence of any influence from Friston or free energy; it debuted in a paper by Rajesh Rao and Dana Ballard in 1999. However, free energy fans have since eagerly explored it.

example, auditory information comes in from the ears, gets relayed first through regions in the brainstem and midbrain, and then goes on to be passed sequentially through several areas in the cortex. This *forward* pathway is widely accepted as crucial for how sensory information gets turned into perception, even by researchers who don't hold much stock in the theory of predictive coding.

What makes predictive coding unique is what it claims about the *backward* pathway – connections going from latter areas to earlier ones (say from the second auditory area in the cortex back to the first). In general, scientists have hypothesised many different roles for these projections. According to the predictive coding hypothesis, these connections carry predictions. For example, when listening to your favourite song, your auditory system may have very precise knowledge about the upcoming notes and lyrics. Under a predictive coding model, these predictions would be sent backwards and get combined with forward-coming information about what's actually happening in the world. By comparing these two streams, the brain can calculate the error between its prediction and reality. In fact, in most predictive coding models, specific 'error' neurons are tasked with just this calculation. Their activity thus indicates just how wrong the brain was: if they are firing a lot, the error in prediction was high, if they're quiet, it was low. In this way, the activity of these neurons is a physical instantiation of free energy. And, according to the free energy principle, the brain should aim to make these neurons fire as little as possible.

Do such error neurons exist in sensory pathways? And does the brain learn to quiet them by making better predictions about the world? Scientists have been looking for answers to these questions for many years. A study conducted by researchers at Goethe University Frankfurt, for example, found that some neurons in the auditory system do decrease their firing when an expected sound is heard. Specifically, the researchers trained mice to press a noise-making lever. When the mice heard the expected sound after pressing the lever, their neurons responded less than if that same sound were played at random or if the lever made an unexpected sound. This suggests the mice had a prediction in mind and the neurons in their auditory system were firing more when that prediction was violated. Overall, however, the evidence for predictive coding is mixed. Not all studies that go looking for error neurons find them and, even when they do, these neurons don't always behave exactly as the predictive coding hypothesis would predict.

Making the brain a better predictive machine might seem like the most obvious way of minimising free energy, but it's not the only one. Because free energy is the difference between the brain's prediction and experience, it can also be minimised by controlling experience. Imagine a bird that has grown accustomed to flying around a certain forest; it can predict which trees will be good for nest building, where the best food is found and so on. One day it flies a little beyond its normal range and finds itself in a city. Experiencing tall buildings and traffic for the first time, its ability to predict almost anything about the world around it is low. This big discrepancy between prediction and experience means free energy is high. To bring its free energy back down, the bird could stick around and hope its sensory systems adapt to be able to predict the features of city life. Or it could simply fly back to the forest it came from. The presence of this second option – choosing actions that result in predictable sensory experiences – is what makes the free energy principle a candidate GUT of the brain. Rather than just explaining features of sensory processing, this principle can encompass decisions about behaviour as well.

The free energy principle has indeed been invoked to explain perception, action and everything in between.[*] This includes processes like learning, sleep and attention, as well as disorders like schizophrenia and addiction. It is also argued that the principle can account for the anatomy of neurons and brain areas, along with the details of how they communicate. In fact, Friston hasn't even constrained free energy to the brain. He's argued for it as a guiding principle of all of biology and evolution and even as a way of understanding the fundamentals of physics.

This tendency to try to wrap complex topics into simple packages has been with Friston throughout his life. In a 2018 *Wired* profile, he recalls a thought he had as a teenager: 'There must be a way of understanding everything by starting from nothing … If I'm only allowed to start off with one point in the entire universe, can I derive everything else I need from that?' In

---

[*] When fully splayed out, the tendrils of the free energy principle reach into many of the topics covered in this book. It builds off the notion of a Bayesian brain (Chapter 10), interacts with ideas from information theory (Chapter 7), uses equations from statistical mechanics (Chapters 4 and 5) and explains elements of visual processing (Chapter 6).

Friston's world, the free energy principle is now the nearly nothing that can explain almost everything.

Outside Friston's world, however, the capabilities of the free energy principle aren't always as obvious. Given its grand promises, countless scientists have attempted to understand the ins and outs of Friston's theory. Few (even those who count themselves fans of the principle) consider their attempts wholly successful. It's not necessarily that the equations are too complicated – many of the scientists trying have dedicated their lives to understanding the mathematics of the mind. Rather, how to extrapolate and apply the free energy principle to all the nooks and crannies of brain function requires a type of intuition that seems to run strongest in Friston himself. Without a clear and objective means of interpreting free energy in any particular case, Friston is left to play the role of free energy whisperer, laying out his take on its implications in countless papers, talks, and Monday meetings.

The confusion around the free energy principle likely results from a feature of it that Friston readily acknowledges: it's not falsifiable. Most hypotheses about how the brain functions are falsifiable – that is, they make claims that can be proven wrong through experiments. The free energy principle, however, is more a way of looking at the brain than a strong or specific claim about how it works. As Friston said: 'The free energy principle is what it is—a principle ... there's not much you can do with it, unless you ask whether measurable systems conform to the principle.' In other words, rather than trying to make clean predictions about the brain based on the free energy principle, scientists should instead ask if the principle helps them see things in a new light. Trying to figure out how a bit of the brain works? Ask if it somehow minimises free energy. If that leads to progress, great; if not, that's fine, too. In this way, the free energy principle is meant to at best offer a scaffolding on which to hang facts about the brain. Insofar as it can connect a great many facts, it is grand and somewhat unifying; however – without falsifiability – its status as a theory is more questionable.

★ ★ ★

Numenta is a small tech company based in Redwood City, California. It was founded by Jeff Hawkins, an entrepreneur who previously founded two companies that produced predecessors to the modern smartphone. Numenta, on the other hand, makes software. The company designs data-processing algorithms aimed to help stockbrokers, energy distributors, IT companies and the like identify and track patterns in streams of incoming data. Numenta's main goal, however, is to reverse-engineer the brain.

Even as he weaved his way through an illustrious career in tech, Hawkins always harboured an interest in the brain. Despite never earning a degree in the field himself, he started the Redwood Neuroscience Institute in 2002. The institute would go on to become part of the University of California, Berkeley, and Hawkins would go on to Numenta in 2005. The work of Numenta is based mainly on ideas presented in a 2004 book Hawkins co-authored with Sandra Blakeslee, *On Intelligence*. The book summarises Hawkins' theory about how the neocortex – the thin layer of brain tissue covering the surface of mammalian brains – works to produce sensation, cognition, learning, movement and more. It's a set of ideas that now goes under the name 'The Thousand Brains Theory of Intelligence'.

At the centre of the Thousand Brains Theory is a piece of neuro-architecture known as the cortical column. Cortical columns are small patches of cells, less than the tip of a pencil in diameter and about four times that in length. They're so-named because they form cylinders running from the top of the neocortex through to the bottom, like so many parallel strands of spaghetti. Looking at a column length-wise, it resembles sheets of sediment: the neurons are segregated into six visibly identifiable layers. Neurons in different layers interact with each other by sending connections up or down. Typically, all the neurons in a column perform a similar function; they may all, for example, respond in a similar way to a sensory input. Yet the different layers do seem to serve some different purposes: some layers, for example, get input from other brain areas and other layers send output off.

Vernon Mountcastle, the sensory neuroscientist who first identified these columns in the mid-twentieth century, believed they represented a fundamental anatomical unit of the brain. Though it went against the dogma of the time, Mountcastle saw potential in the idea of a single repeating unit that tiles the whole of the neocortex – a single unit that could process the full variety of information the cortex receives. Hawkins agrees. In his book, he describes Mountcastle's work as 'the Rosetta stone of neuroscience' because it is 'a single idea that united all the diverse and wondrous capabilities of the human mind'.

To understand what Hawkins thinks these mini-processing units do we have to consider both time and space. 'If you accept the fact intelligent machines are going to work on the principles of the neocortex,' Hawkins said in a 2014 interview, '[time] is the entire thing.' Inputs to the brain are constantly changing; this makes a static model of brain function woefully incomplete. What's more, the outputs of the brain – the behaviours produced by the body – are extended through both space and time. According to Hawkins, actively moving the body through space and getting dynamic streams of sensory data in return helps the brain build a deep understanding of the world.

Neuroscientists know a bit about how animals move through the world. It has a lot to do with a type of neuron called a 'grid cell' (see Figure 26).* Grid cells are neurons that are active when an animal is in certain special locations. Imagine a mouse running around an open field. One of its grid cells will be active when the mouse is right in the middle of the field. That same cell will *also* be active when the mouse has moved a few body lengths north of the centre and then again a few lengths north of that. The same pattern of activity would also be seen if the mouse went 60 degrees west of north instead. In fact, if you made a map of all the places this cell is active it would form a polka-dot pattern across the

---

* Edvard Moser, May-Britt Moser and John O'Keefe were awarded the Nobel Prize in 2014 for their discovery of these cells, along with the closely related, but more obviously named, 'place cells'.

whole field. These polka dots would all be evenly spaced at the vertices of a triangular grid (hence the name). Different grid cells differ in the size and orientation of this grid, but they all share this common feature.

Impressed by their ability to represent space, Hawkins made grid cells an integral part of his theory of how the neocortex learns about the world. There is a problem, however: grid cells aren't found in the neocortex. Instead, they reside in an evolutionarily older part of the brain known as the entorhinal cortex. Despite little evidence for grid cells outside of this region, Hawkins hypothesises that they are actually hiding away in the sixth layer of every column in the neocortex.

What exactly are they doing there? To explain this, Hawkins likes to use the example of running your finger around a coffee cup (Hawkins actually attributes the origins of his theory to a eureka moment he had while contemplating a coffee cup and will even bring the cup to talks for demonstration). Columns in the sensory processing part of the neocortex will get inputs from the fingertip. According to Hawkins' theory, the grid cells at the bottom of these columns will also be tracking the location of the fingertip. Combining information about

Mouse explores an environment:

The firing locations of two example grid cells in this environment:

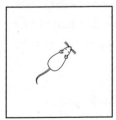 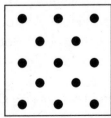 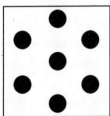

Figure 26

where the finger is and what the cup feels like there, the column can learn the shape of the object as it navigates around it. The next time the same object is encountered the column can use this stored knowledge to recognise it.

As these cortical columns exist across the neocortex, this process could be happening in parallel everywhere. The columns that represent other parts of the hand, for example, would be building their own models of the coffee cup as they come into contact with it. And areas in the visual system would combine visual information with the location of the eyes to build their own understanding of the cup, too. In total, a cohesive understanding of the world is built through these distributed contributions, like thousands of brains working in unison.

Hawkins' theory is ever-evolving and many of the details are still to be worked out, but his hopes for it are high. The way he sees it, just as columns could learn the shapes of literal objects, so too could they learn the shapes of abstract ones. Navigating the space of thought or language can be accomplished with the same mechanisms for navigating the real, physical world. If this is true, it explains how a repeating pattern in the neocortex can be used to do so many different things, from vision to audition, movement to mathematics.

Exactly how identical these columns are, however, is a subject of debate. At a glance the neocortex may appear like a uniform tessellation, but upon closer inspection differences do emerge. Some studies have found that the size of columns, the number and type of neurons they contain, and how they interact with each other varies across regions of the neocortex. If these columns are not actually anatomically identical, they may differ in function, too. This means that each brain region could be more specialised for the kind of tasks it has to perform than the Thousand Brains Theory assumes. If that's the case, the hope for a universal algorithm of intelligence may be dashed.

As seen throughout this book, mathematical models of the brain are usually built by first identifying – out of the mounds of available data – a selection of facts that seem relevant. Those facts are then simplified and pieced together in a way that demonstrates how, in theory, a bit of the brain could work. Additionally, in figuring out how exactly to cobble this toy version of biology together, some new and surprising predictions may be made. When compared to this schema, the Thousand Brains Theory is a model like any other in neuroscience. Indeed, many of the component concepts – columns, grid cells, object recognition – have already been studied extensively by means of both experimental and computational work. In this way, the theory is not unique; it's a guess that may be right, may be wrong or may – like many theories – be a little of both.

Perhaps what sets the work of Hawkins and Numenta apart then is simply the unrelenting optimism that this one *is* different – that for the first time a key that will unlock all the doors of the cortex is truly within reach. When asked in a 2019 interview about how far off a full understanding of the neocortex is, Hawkins said: 'I feel I've already passed a step function. So, if I can do my job correctly over the next five years – meaning I can proselytise these ideas, I can convince other people they're right, we can show that other machine learning people should pay attention to these ideas – then we're definitely in an under 20-year timeframe.' That's a confidence not commonly seen in scientists – and because Hawkins has the private funding to back it, it's a confidence that is not constrained by normal scientific pressures.

Hawkins is known to be self-assured in his claims about the brain. However, his past ability to deliver on promises for breakthrough brain-based algorithms has been questionable. Geoffrey Hinton, a leader in artificial intelligence research, has described Hawkins' contributions to the field as 'disappointing'. And in 2015, psychology professor Gary Marcus compared Numenta's work to other AI techniques by saying: 'I have not seen a knock-down argument that they yield better performance in any major challenge area.' What chance does the Thousand Brains Theory have of ever providing the field with a set of truly universal mechanisms of intelligence? Only time – a concept so central to Hawkins' thinking – will tell.

★ ★ ★

By some accounts, no theory of the brain could be complete without explaining its biggest and most enduring mystery: consciousness. The C-word can be a tough subject for scientists, laden as it is with centuries of philosophical baggage. Yet, in the eyes of certain researchers, a precise scientific definition that could be used to not just identify but also *quantify* consciousness anywhere it may exist is the holy grail of their work. It is also the promise of 'integrated information theory'.

Integrated information theory (or IIT) is an attempt to define consciousness with an equation. It was originally put forth by Italian neuroscientist Giulio Tononi in 2004 and has been iterated on by him and others ever since. IIT is designed to measure consciousness in anything: in computers, rocks and aliens as easily as in brains. By making a universal claim of what consciousness is, it differs from the more biology-centred theories devised by some neuroscientists.

IIT is able to free itself from the specific physical features of the brain because its inspiration comes from another source entirely: introspection. By reflecting on the first-person conscious experience, Tononi came up with five important traits fundamental to consciousness; these are the 'axioms' on which IIT is built. The first axiom is the basic fact that consciousness exists. Others include the observation that a conscious experience is composed of multiple distinct sensations, the experience is specific, it appears to us as an integrated whole and it is uniquely what it is – no more or less.

Tononi considered what kinds of information-processing systems could give rise to these axioms of experience. Through this, he was able to map the axioms to mathematical terms. The end result is a unified measure of what is called 'integrated information', a value Tononi symbolises with the Greek letter *phi*. In total, phi indicates just how intermixed the information in a system is. The right kind of intermixing is supposed to give rise to the richness and wholeness of experience. According to IIT, the higher the phi that a system has, the more conscious it is.

As it turns out, calculating the phi for a system with any reasonable amount of complexity is nearly impossible. For the human brain, it would first require conducting a near endless amount of experiments in order to probe how the different sub-structures of the brain interact. Even if that could be done, a long and gruelling series of computations would then begin. To overcome this hurdle, multiple approximations to phi have been devised. Through this, it is possible to make educated guesses about the phi in a system. This has been used to explain why certain brain states lead to more conscious experience than others. For example, during sleep, the ability of neurons to communicate effectively is interrupted. This makes the brain less able to integrate information, resulting in lower phi. According to Tononi's theory, similar reasoning can explain the unconsciousness that comes with seizures as well.

The theory also makes some, perhaps surprising, predictions. For example, the phi of an average thermostat is small, but still not zero. This implies that the device regulating your room temperature has some amount of conscious experience. What's more, some very simple devices – if built just right – can actually have a value of phi much higher than the estimated phi of the human brain. These counterintuitive conclusions make some scientists and philosophers sceptical of IIT.

Another critique of the theory is targeted at its axiomatic basis. According to this argument, the axioms Tononi chose aren't the only ones that a theory of consciousness could be built on. And his way of mapping these axioms to mathematics isn't obviously the only, or best, way either. The problem is: if the foundations of IIT are arbitrary, then how can we trust the conclusions that spring from them, especially when they surprise us?

An informal survey of consciousness scientists conducted in 2018 revealed that IIT was not the favoured theory among experts (it came in fourth after two other theories and the catch-all category of 'other'). But the same survey found that IIT fared better among non-experts: in fact, it was rated first among the subset of non-experts who felt they had enough knowledge to respond. Some of the survey authors suspect this may be a result of IIT's PR. From the outside, the theory looks well founded if only because it has the authority of hard math behind it. And more than most scientific theories of consciousness, IIT has been featured in the popular press. This includes writings by Christof Koch, a prominent neuroscientist who has become a collaborator of Tononi's and a public advocate of IIT. In his book, *Consciousness: Confessions of a Romantic Reductionist*, Koch describes his personal journey through the scientific study of consciousness, including work he did with Nobel Prize winner Francis Crick, and his views on IIT.* Such popular accounts may be effective in getting the theory out to a broader audience, but don't necessarily convince scientists in the know.

Even scientists who lack faith in the power of IIT still tend to applaud the attempt. Consciousness is a notoriously

---

* Tononi himself has also written a book aiming to explain his theory to a broader audience. In *Phi: A Voyage from the Brain to the Soul*, Tononi tells a fictional tale of seventeenth-century scientist Galileo Galilei exploring notions of consciousness through interactions with characters inspired by Charles Darwin, Alan Turing and Crick.

difficult concept to tame, which makes IIT's effort to submit it to orderly scientific inquiry still a step in the right direction. As vocal critic of IIT physicist Scott Aaronson wrote on his blog: 'The fact that integrated information theory is wrong – demonstrably wrong, for reasons that go to its core – puts it in something like the top 2 per cent of all mathematical theories of consciousness ever proposed. Almost all competing theories of consciousness, it seems to me, have been so vague, fluffy and malleable that they can only **aspire** to wrongness.'

★ ★ ★

GUTs can be a slippery thing. To be grand and unifying, they must make simple claims about an incredibly complex object. Almost any statement about 'the brain' is guaranteed to have exceptions lurking somewhere. Therefore, making a GUT too grand means it won't actually be able to explain much specific data. But, tie it too much to specific data and it's no longer grand. Whether untestable, untested, or tested and failed, GUTs of the brain, in trying to explain too much, risk explaining nothing at all.

While this presents an uphill battle for GUT-seeking neuroscientists, it's less of a challenge in physics. The reason for this difference may be simple: evolution. Nervous systems evolved over eons to suit the needs of a series of specific animals in specific locations facing specific challenges. When studying such a product of natural selection, scientists aren't entitled to simplicity. Biology took whatever route it needed to create functioning organisms, without regard to how understandable any part of them would be. It should be no surprise, then, to find that the brain is a mere hodgepodge of different components and mechanisms. That's all it needs to be to function. In total, there is no guarantee – and maybe not even any compelling reasons to expect – that the brain can be described by simple laws.

Some scientists choose to embrace this mess. Instead of reducing the brain to its barest elements, they build a form of 'grand unified model' that sticks all of the parts together. While traditional GUTs have the simplicity of a steak cooked with just a sprinkle of salt, these models are more like a big pot of soup. And while they're not as sleek and elegant as GUTs, they may be better equipped to get the job done.

The hyper-detailed simulation built by the Blue Brain Project, as discussed in Chapter 2, is one example of this more inclusive approach. These researchers extracted countless details about neurons and synapses through a series of painstaking experiments. They then pieced all this data back together into an elaborate computational model of just a small speck of the brain. Such an approach assumes that each detail is precious and that the brain won't be understood by stripping them away. It's a wholehearted embrace of the nuance of biology, with the hope that by throwing everything together, a fuller understanding of what makes the brain work will emerge. The trouble here, however, is scale. A bottom-up approach to rebuilding the brain can only proceed one neuron at a time, which means a complete model is a long way off.

The Semantic Pointer Architecture Unified Network, better known as SPAUN, comes at things from a very different direction. Rather than capturing all the minutiae of the neurobiology, SPAUN – developed by a team working under Chris Eliasmith at University of Waterloo in Ontario, Canada – is about making a model of the brain that works. That means getting the same sensory inputs and having the same motor outputs. Specifically, SPAUN has access to images as input and controls a simulated arm to write its outputs. In between this input and output is a complex web of 2.5 million simple model neurons, arranged to broadly mimic the structure of the whole brain. Through these neural connections SPAUN can perform seven different cognitive and motor tasks, such as drawing digits, recalling lists of objects and completing simple patterns. In this way, SPAUN eschews elegance for function. Of course, the human brain contains tens of thousands of times as many neurons and can do a lot more than seven tasks. Whether the principles of utility and scale that got SPAUN to where it is can take it all the way up to a full model of the brain – or whether more of the nuances of neurons need to be added – is unknown.

True GUTs aim to condense. They melt diverse information down into a compact and digestible form. This makes GUTs seem satisfying because they give the sense that the workings of the brain can be fully grasped in one hand. Models like SPAUN and the Blue Brain Project simulation, however, are expansive. They bring in many sources of data and use them to build an elaborate structure. In this way, they sacrifice interpretability for accuracy. They aim to explain everything by incorporating everything there is to explain.

Though as with all models, even these more expansive ones are still not perfect replicas. Makers of these models still need to choose what to include and what not to include, what they aim to explain and what they can ignore. When aiming for something akin to a GUT, the hope is always to find the simplest set of principles that can explain the largest set of facts. With an object as dense and messy as the brain, that simple set may still be pretty complicated. To know in advance what level of detail and what magnitude of scale will be needed to capture the relevant features of brain function is impossible. It is only through the building and testing of models that progress on that question can be made.

On the whole, neuroscience has enjoyed a very fruitful relationship with the 'harder', more quantitative sciences. It has received many gifts from the likes of physics, mathematics and engineering. These analogies, methods and tools have shifted thinking about everything from neurons to behaviour. And the study of the brain has given back in return, providing inspiration for artificial intelligence and a testing ground for mathematical techniques.

But neuroscience is not physics. It must avoid playing the role of the kid sibling, trying to follow exactly in the footsteps of this older discipline. The principles that guide physics and the strategies that have led it to success won't always work when applied to biology. Inspiration, therefore, must be taken with care. When building models of the mind, the aesthetics of the mathematics is not the only guiding light. Rather, this influence needs always to be weighed against the unique realities of the brain. When balanced just right, the intricacies of the biology can be reduced to mathematics in a way that produces true insights and isn't overly influenced by other fields. In this way, the study of the brain is forging its own path for how to use mathematics to understand the natural world.

Mathematical Appendix

Chapter 2: How Neurons Get Their Spike

Lapicque devised an equation for describing how the voltage across a cell's membrane changes over time. The equation is based on those used to describe electrical circuits. Specifically, the voltage, $V(t)$, is defined according to the equation for a circuit with resistance (R) and capacitance (C) in parallel:

$$\tau \frac{dV}{dt} = -(V(t) - V_{rest}) + RI(t)$$

where $\tau = RC$. External input to the cell (from an experimenter or the effects of another neuron) is represented by $I(t)$. The cell membrane thus integrates this external input current, with some leak.

Lapicque's equation doesn't capture what happens to the membrane potential during an action potential. However, we can add a simple mechanism that indicates when the cell's membrane has reached its threshold and would cause a spike. Specifically, to turn this equation into a model of a neuron that fires, the voltage is reset to its resting state (V_{rest}) once it reaches a spiking threshold (V_{thresh}):

$$V(t) = V_{rest}, \text{ if } V(t) = V_{thresh}$$

This doesn't mimic the complex dynamics of the action potential (for which the Hodgkin-Huxley model is needed), but it does provide a simple way to calculate spike times.

Chapter 3: Learning to Compute

The perceptron is a one-layer artificial neural network that can learn to perform simple classification tasks. The learning occurs via updates in the weights between the input neurons and the output neuron, calculated based on specific examples of inputs and outputs.

The learning algorithm starts with a set of random weights, w_n, one for each of the N binary inputs, x_n. The output classification, y, of the perceptron is calculated as:

$$y(\boldsymbol{x}) = \begin{cases} 1, & \text{if } \sum_{n=1}^{N} w_n x_n + b \geq 0 \\ 0, & \text{otherwise} \end{cases}$$

where b is a bias that shifts the threshold. With learning, each entry of **w** updates according to the learning rule:

$$w_n \leftarrow w_n + \lambda(y^* - y(\boldsymbol{x}))x_n$$

where y^* is the correct classification and λ is the learning rate. If x_n is one, the sign of the difference between the correct classification and the perceptron's output will determine how w_n is updated. If x_n or the difference is zero, no update occurs.

Chapter 4: Making and Maintaining Memories

A Hopfield network represents memories as patterns of neural activity. The connections between the neurons allow the network to implement associative memory – that is, a full memory can be retrieved by activating a subset of it.

The network is composed of N cells, the interactions between which are defined according to a symmetric weight matrix (W). Each entry (w_{nm}) in this matrix defines the strength of the connection between cells n and m. At each time point, the activity state for each cell (c_n for $n=1...N$) is updated according to:

$$c_n \leftarrow \begin{cases} 1, & \text{if } \sum_{m=1}^{N} w_{nm}c_m \geq \theta_n \\ -1, & \text{otherwise} \end{cases}$$

where θ_n is a threshold.

Each memory, ϵ^i, is a length N vector defining the activity state of each neuron. If the activity of the network is initially set to a noisy, partial version of a memory, it will evolve to that memory's attractor state (defined by ϵ^i), at which point the network activity $\mathbf{c}$ will stop changing.

The weight matrix is determined by the memories stored in the network. To store K memories, each entry of W is defined according to:

$$w_{nm} = \frac{1}{K} \sum_{i=1}^{K} \epsilon_n^i \epsilon_m^i$$

Therefore, pairs of neurons that have similar activity in many memories will have strong positive connections; those with opposite activity patterns will have strong negative connections.

Chapter 5: Excitation and Inhibition

Networks with the appropriate balance between excitation and inhibition can create stable, noisy neural activity. These networks can be analysed using the mean-field approach, which simplifies the mathematics of the full network into just a handful of equations.

The mean-field equations for the balanced network start with a network of N neurons (both excitatory and inhibitory) wherein neurons receive external input as well as recurrent input. For the recurrent input, each neuron receives K excitatory and K inhibitory inputs. K is assumed to be much less than N:

$$1 << K << N$$

Looking at the case of large K and a constant external input to the network, the mean input to a cell of type j (either excitatory or inhibitory) is given by:

$$\mu_j = \sqrt{K}(X_j m_X + m_E - W_{jI} m_I) - \theta_j$$

And the variance of this input is:

$$\sigma_j^2 = m_E + W_{jI}^2 m_I$$

The terms X_j and m_x represent the external input's connection strength to the j population and its firing rate respectively; θ_j is the threshold for spiking. W_{jI} is a measure of the total strength from the inhibitory population to the j population (the corresponding value from the excitatory population is defined to be one). W_{jI} is given as the strength of a single connection times $\sqrt{K}$.

m_j is the average activity of the j population defined to range from zero to one. These values are determined by the mean and square root of the variance of the input according to:

$$m_j = H\left(\frac{-\mu_j}{\sqrt{\sigma_j^2}}\right)$$

where H is the complementary error function.

To make sure that neither the excitatory nor inhibitory input to a cell overwhelms its output, the first term in the equation for μ_j should be of the same order as the threshold, which is one. To satisfy this, individual connections should have a strength of $1/\sqrt{K}$.

Chapter 6: Stages of Sight

Convolutional neural networks process images by replicating some of the basic features of the brain's visual system. They are composed of several basic operations. Starting with an image I, the first step is to convolve this image with a filter F. The result of this convolution is passed through an elementwise nonlinearity (ϕ) to yield activity for the simple cell-like layer:

$$A_s = \phi(I * F)$$

The most common nonlinearity is positive rectification:

$$\phi(x) = \max(x, 0)$$

Assuming the image and filter are both two-dimensional matrices, A_s is also a two-dimensional matrix. To replicate complex cell responses, a 2D max-pooling operation is applied to the simple cell-like activity. Each element of the matrix of complex cell-like activity (A^c) is defined according to:

$$a_{ij}^c = \max_{pq \in Pij} a_{pq}^s$$

where P_{ij} is a 2D neighbourhood of A_s centred on location ij. The effect of this operation is that the complex cell activity is simply the maximal activity of the patch of simple cells from which it receives inputs.

Chapter 7: Cracking the Neural Code

Shannon defined information in terms of bits, which are calculated as the base-two log of a symbol's inverse probability. This can also be written as the negative of the base-two log of the probability:

$$\log_2 P(\frac{1}{x_i}) = -\log_2 P(x_i)$$

The total information in a code, a value known as entropy (H), is a function of the information in each of its symbols. Specifically, entropy is a sum of the information contained in each symbol (x_i) of a code X weighted by its probability, $P(x_i)$.

$$H(X) = -\sum_i P(x_i) \, \log_2 P(x_i)$$

Chapter 8: Movement in Low Dimensions

Principal components analysis (PCA) can be used to reduce the dimensionality of the activity of a population of neurons. Applying PCA to neural data starts with a matrix of data (X) wherein each row represents a neuron (out of a total of N neurons) and each column is the mean-subtracted activity of those neurons over time (of length L):

$$X \in \mathbb{R}^{N \times L}$$

The covariance matrix of this data is given as

$$K = XX^T$$

The eigenvalue decomposition says:

$$K = Q\Lambda Q^{-1}$$

where each column in Q is an eigenvector of K and Λ is a diagonal matrix wherein the entries on the diagonal are the eigenvalues of the corresponding eigenvectors. The principal components of the data are defined as the eigenvectors of K.

In order to reduce the full-dimensional data to D dimensions, the top D eigenvectors (as ranked by their eigenvalues) are used as the new axes. Projecting the full-dimensional data on to these new axes provides a new data matrix:

$$X_{reduced} \in \mathbb{R}^{D \times L}$$

If D is three or less, this reduced data matrix can be visualised.

Chapter 9: From Structure to Function

Watts and Strogatz argued that many real-world graphs could be described as small-world networks. Small-world networks have low average path lengths (the number of edges traversed between any two nodes) and high clustering coefficients.

Assume a graph composed of N nodes. If a given node n is connected to k_n other nodes (known as its neighbours), then the clustering coefficient of that node is:

$$c_n = \frac{E_n}{k_n(k_n - 1)/2}$$

where E_n is the number of edges that exist amongst n's neighbours and the term in the denominator is the total number of edges that could exist among these nodes. The clustering coefficient is thus a measure of how interconnected, or 'cliquey', groups of nodes are.

The clustering coefficient for the entire network is given by the average of the clustering coefficients for each node:

$$C = \frac{1}{N} \sum_{n=1}^{N} c_n$$

Chapter 10: Making Rational Decisions

The full form of Bayes' rule is:

$$P(h \mid d) = \frac{P(d \mid h)P(h)}{P(d)}$$

where h represents the hypothesis and d the observed data. The term on the left-hand side of the equation is known as the posterior distribution. Bayesian decision theory (BDT) addresses how Bayes' rule can guide decisions by indicating how the posterior distribution should be mapped to a specific perception, choice or action.

In BDT, a loss function indicates the penalty that comes from making different types of wrong decision (for example, incorrectly seeing a red flower as white versus seeing a white flower as red could have different negative outcomes). In the most basic loss function, any incorrectly chosen hypothesis incurs the same penalty while the correct choice (h^*) has no penalty:

$$l(\hat{h}, h^*) = \begin{cases} 1, \text{if } \hat{h} \neq h^* \\ 0, \text{otherwise} \end{cases}$$

The overall expected loss for choosing a certain hypothesis (h) is calculated by weighing this loss by the probability of each hypothesis:

$$L(\hat{h}) = \sum_h l(\hat{h}, h)P(h \mid d)$$

which yields:

$$L(\hat{h}) = 1 - P(h = \hat{h} \mid d)$$

Therefore, to minimise this loss, the option that maximises the posterior distribution should be chosen. That is, the best hypothesis is the one with the highest posterior probability.

Chapter 11: How Rewards Guide Actions

Reinforcement learning describes how animals or artificial agents can learn to behave simply by receiving rewards. A central concept in reinforcement learning is value, a measure that combines the amount of reward received currently with what is expected to come in the future.

The Bellman equation defines the value (V) of a state (s) in terms of the reward (R) received if action a is taken in that state and the discounted value of the next state:

$$V(s) = \max_a \left[R(s, a) + \beta V(T(s, a)) \right]$$

Here, β is the discounting factor and T is the transition function, which determines which state the agent will be in after taking action a in state s. The max operation functions to ensure that the action that yields the highest value is always taken. You can see that the definition of value is recursive as the value function itself appears on the right-hand side of the equation.

Chapter 12: Grand Unified Theories of the Brain

The free energy principle has been offered as a unifying theory of the brain that can describe neural activity and behaviour. Free energy is defined as:

$$F(s, \mu) = -\log p(s) + D_{KL} \left[q(x \mid \mu) \mid\mid p(x \mid s) \right]$$

where s are sensory inputs, μ are internal brain states and x are states of the world. The first term in this definition (the negative log probability of s) is sometimes referred to as 'surprise', as it is high when the probability of the sensory inputs is low.

D_{KL} is the Kullback–Leibler divergence between two probability distributions, defined as:

$$D_{KL}\left[q \mid\mid p\right] = \sum_{y \in Y} q(y) \log \frac{q(y)}{p(y)}$$

The second term of the definition of free energy thus measures the difference between the probability of states of the world given the brain's internal state and the probability of states of the world given the sensory inputs. The brain can be thought of as attempting to approximate $p(x \mid s)$ using its own internal states $(q(x \mid \mu))$ and the better the approximation the lower the free energy.

Because the free energy principle says the brain aims to minimise free energy, it should update its internal states according to:

$$\mu = \min_{\mu} F(s, \mu)$$

In addition, the choice of actions (a) taken by the animal will influence the sensory inputs it receives:

$$s' = f(a)$$

Therefore, actions should also be selected according to their ability to minimise free energy:

$$a = \min_{a} F(s', \mu')$$

Acknowledgements

As I write this I am pregnant with my first child. They say it takes a village to raise a kid. I believe that is eventually true, but so far it has been a relatively solitary experience. Writing a book, on the other hand, seems to take a village from the start.

First, I have to thank my husband, Josh. We met while both getting our PhDs at the Center for Theoretical Neuroscience at Columbia, which meant he had to serve both as moral support and as fact-checker throughout this process. He also did a good job of making sure I at least occasionally ate proper meals and saw some friends. I'd also like to thank members of his family – Sharon, Roger, Laurie – for their support and excitement.

I owe a tremendous debt to the NeuWrite community. I first joined this group of scientists and writers when I was a graduate student in New York and quickly joined the London chapter once I moved to the UK. In addition to putting me in contact with the people at Sigma, the members of NeuWrite also provided advice and commiseration about book-writing in general. The opportunity to regularly workshop my chapters with this group helped quell my writing anxiety and, of course, made the book better. Special thanks to Liam Drew, Helen Scales, Roma Agrawal and Emma Bryce.

I've relied on many friends – both practising neuroscientists and not – to provide input and feedback on the book. They've made these chapters clearer. Much thanks to Nancy Padilla, Yul Kang, Vishal Soni, Jessica Obeysekare, Victor Pope, Sarahjane Tierney, Jana Quinn, Jessica Graves, Alex Cayco-Gajic, Yann Sweeney and (my sister) Ann Lindsay.

I also relied on the diverse and amorphous community of researchers known as 'neuroscience twitter' to sound ideas off and crowdsource resources. Many thanks to this engaged crowd of friends and strangers alike!

I reached out to some researchers with particular expertise to look over different chapters. I'm very thankful for the time and knowledge of Athanasia Papoutsi, Richard Golden, Stefano Fusi, Henning Sprekeler, Corey Maley, Mark Humphries, Jan Drugowitsch and Blake Richards. Of course, any errors that remain in the text are my own fault.

The Bloomsbury Sigma team is the reason this book is a reality instead of merely an ill-formed wish in my mind. Thanks to Jim Martin, Angelique Neumann and Anna MacDiarmid for shepherding both me and the book through the process.

A general thanks is due to my family (particularly my sisters, Sara and Ann) and friends who have been patiently hearing about 'the book' for some time. And finally, I'd like to express my gratitude to the computational neuroscience community as a whole. Floating around this field for nearly 10 years and soaking in knowledge from many different researchers gave me the proper foundation for writing this book.

Bibliography

Chapter 1: Spherical Cows

Abbott, L. F. 2008. Theoretical neuroscience rising. *Neuron* 60(3):489–95 doi:10.1016/j.neuron.2008.10.019.

Cajal, S. R. 2004. *Advice for a Young Investigator*. MIT Press, Massachusetts, USA.

Lazebnik, Y. 2002. Can a biologist fix a radio Or, what I learned while studying apoptosis. *Cancer Cell* 2(3):179–82 doi:10.1016/s1535-6108(02)00133-2.

Nakata, K. 2013. Spatial learning affects thread tension control in orb-web spiders. *Biology Letters* 9(4) doi:10.1098/rsbl.2013.0052.

Russell, B. 2009. *The Philosophy of Logical Atomism*. Routledge, London.

Chapter 2: How Neurons Get Their Spike

Branco, T., *et al.* 2010. Dendritic discrimination of temporal input sequences in cortical neurons. *Science* 329(5999):1671–75 doi:10.1126/science.1189664.

Bresadola, M. 1998. Medicine and science in the life of Luigi Galvani (1737–98). *Brain Research Bulletin* 46(5):367–80 doi:10.1016/s0361-9230(98)00023-9.

Brunel, N. & van Rossum, M. C. W. 2007. Lapicque's 1907 paper: From frogs to integrate-and-fire. *Biological Cybernetics*, 9(5):337–39 doi:10.1007/s00422-007-0190-0.

Burke, R. E. 2006. John Eccles' pioneering role in understanding central synaptic transmission. *Progress in Neurobiology* 78(3):173–88 doi:10.1016/j.pneurobio.2006.02.002.

Cajori, F. 1962. *History of Physics*. Dover Publications, New York, USA.

Finkelstein, G. 2013. *Emil Du Bois-Reymond: Neuroscience, Self, and Society in Nineteenth-Century Germany*. MIT Press, Massachusetts, USA.

Finkelstein, G. 2003. M. Du Bois-Reymond goes to Paris. *The British Journal for the History of Science* 36(3):261–300 www.jstor.org/stable/4028156. JSTOR.

Volta, A. & Banks, J. 1800. On the electricity excited by the mere contact of conducting substances of different kinds. *The Philosophical Magazine* 7(28):289–311 doi:10.1080/14786440008562590.

Huxley, A. F. 1964. Excitation and conduction in nerves: quantitative analysis. *Science* 145(3637):1154–59 doi:10.1126/science.145.3637.1154.

Bynum, W. F. & Porter, R. 2006. Johannes Peter Müller. *Oxford Dictionary of Scientific Quotations*. OUP, Oxford.

Kumar, A., *et al.* 2011. The role of inhibition in generating and controlling parkinson's disease oscillations in the basal ganglia. *Frontiers in Systems Neuroscience* 5 doi:10.3389/fnsys.2011.00086.

Tyndall, J. 1876. Lessons in electricity IV. *Popular Science Monthly* 9. Wikisource.

Markram, H., *et al.* 2015. Reconstruction and simulation of neocortical microcircuitry. *Cell* 163(2):456–92 doi:10.1016/j.cell.2015.09.029.

McComas, A. 2001. *Galvani's Spark: The Story of the Nerve Impulse.* Oxford University Press, USA.

Piccolino, M. 1998. Animal electricity and the birth of electrophysiology: the legacy of Luigi Galvani. *Brain Research Bulletin* 46(5):381–407 doi:10.1016/s0361-9230(98)00026-4.

Schuetze, S. M. 1983. The discovery of the action potential. *Trends in Neurosciences* 6:164–68 doi:10.1016/0166-2236(83)90078-4.

Squire, L. R., editor. 1998. *The History of Neuroscience in Autobiography, Volume 1.* Academic Press, Cambridge, Massachusetts, USA.

Squire, L. R., editor. 2003. *The History of Neuroscience in Autobiography, Volume 4.* Academic Press, Cambridge, Massachusetts, USA.

Squire, L. R., editor. 2006. *The History of Neuroscience in Autobiography, Volume 5.* Academic Press, Cambridge, Massachusetts, USA.

Chapter 3: Learning to Compute

Le, Q. V. & Schuster, M. 2016. A neural network for machine translation, at production scale. Google AI Blog. ai.googleblog.com/2016/09/a-neural-network-for-machine.html. Accessed 13 April 2020.

Albus, J. S. 1971. A theory of cerebellar function. *Mathematical Biosciences* 10(1):25–61 doi:10.1016/0025-5564(71)90051-4.

Anderson, J. A. & Rosenfeld, Edward. 2000. *Talking Nets: An Oral History of Neural Networks*. MIT Press, Massachusetts, USA.

Arbib, M. A. 2000. Warren McCulloch's search for the logic of the nervous system. *Perspectives in Biology and Medicine* 43(2):193–216 doi:10.1353/pbm.2000.0001.

Bishop, G. H. 1946. Nerve and synaptic conduction. *Annual Review of Physiology* 8:355–74 doi:10.1146/annurev.ph.08.030146.002035.

Garcia, K. S., *et al.* 1999. Cerebellar cortex lesions prevent acquisition of conditioned eyelid responses. *Journal of Neuroscience* 19(24):10940–47 doi:10.1523/JNEUROSCI.19-24-10940.1999.

Gefter, A. 2015. The man who tried to redeem the world with logic. *Nautilus* http://nautil.us/issue/21/information/the-man-who-tried-to-redeem-the-world-with-logic.

Hartell, N. A. 2002. Parallel fiber plasticity. *Cerebellum* 1(1):3–18 doi:10.1080/147342202753203041.

Linsky, B. & Irvine, A. D. 2019. Principia Mathematica. *The Stanford Encyclopedia of Philosophy*, edited by Zalta, E. N., Metaphysics Research Lab, Stanford University https://plato.stanford.edu/archives/fall2019/entries/principia-mathematica/.

McCulloch, W. S. 2016. *Embodiments of Mind*. MIT Press, Massachusetts, USA.

Papert, S. 1988. One AI or many? *Daedalus* 117(1):1–14 www.jstor.org/stable/20025136. JSTOR.

Piccinini, G. 2004. The first computational theory of mind and brain: a close look at McCulloch and Pitts's logical calculus of ideas immanent in nervous activity. *Synthese* 141(2):175–215 doi:10.1023/B:SYNT.0000043018.52445.3e.

Rosenblatt, F. 1957. *The Perceptron, a Perceiving and Recognizing Automaton Project Para*. Cornell Aeronautical Laboratory, New York, USA.

Russell, B. 2014. *The Autobiography of Bertrand Russell*. Routledge, London.

Schmidhuber, J. 2015. Who invented backpropagation? http://people.idsia.ch/~juergen/who-invented-backpropagation.html. Accessed 13 April 2020.

Chapter 4: Making and Maintaining Memories

Bogacz, R., *et al.* 2001. A familiarity discrimination algorithm inspired by computations of the perirhinal cortex. *Emergent Neural Computational Architectures Based on Neuroscience: Towards Neuroscience-Inspired Computing.* Springer-Verlag, Switzerland 428–441.

Brown, R. E. & Milner, P. M. 2003. The legacy of Donald O. Hebb: more than the Hebb synapse. *Nature Reviews Neuroscience* 4(12):1013–19 doi:10.1038/nrn1257.

Chumbley, J. R., *et al.* 2008. Attractor models of working memory and their modulation by reward. *Biological Cybernetics* 98(1):11–18 doi:10.1007/s00422-007-0202-0.

Cooper, S. J. 2005. Donald O. Hebb's synapse and learning rule: a history and commentary. *Neuroscience and Biobehavioral Reviews* 28(8):851–74 doi:10.1016/j.neubiorev.2004.09.009.

Fukuda, K., *et al.* 2010. Discrete capacity limits in visual working memory. *Current Opinion in Neurobiology* 20(2):177–82 doi:10.1016/j.conb.2010.03.005.

Fuster, J. M. & Alexander, G. E. 1971. Neuron activity related to short-term memory. *Science* (New York, USA) 173(3997):652–54 doi:10.1126/science.173.3997.652.

Hopfield, J. J. 2014. Whatever happened to solid state physics? *Annual Review of Condensed Matter Physics* 5(1):1–13 doi:10.1146/annurev-conmatphys-031113-133924.

Hopfield, J. J. 2018. Now what? Princeton Neuroscience Institute https://pni.princeton.edu/john-hopfield/john-j.-hopfield-now-what. Accessed 13 April 2020.

Kim, Sung Soo, *et al.* 2017. Ring attractor dynamics in the *Drosophila* central brain. *Science* (New York, USA) 356(6340):849–53 doi:10.1126/science.aal4835.

Lechner, H. A., *et al.* 1999. 100 years of consolidation – remembering Müller and Pilzecker. *Learning & Memory* 6,(2):77–87 doi:10.1101/lm.6.2.77.

MacKay, D. J. C. 2003. *Information Theory, Inference and Learning Algorithms.* Cambridge University Press, UK.

Martin, S. J. & Morris, R. G. M. 2002. New life in an old idea: the synaptic plasticity and memory hypothesis revisited. *Hippocampus* 12(5):609–36 doi:10.1002/hipo.10107.

Pasternak, T. & Greenlee, M. W. 2005. Working memory in primate sensory systems. *Nature Reviews Neuroscience* 6(2):97–107 doi:10.1038/nrn1603.

Zhang, K. 1996. Representation of spatial orientation by the intrinsic dynamics of the head-direction cell ensemble: a theory. *Journal of Neuroscience.* www.jneurosci.org/content/16/6/2112. Accessed 13 April 2020.

Roberts, A. C. & Glanzman, D. L. 2003. Learning in aplysia: looking at synaptic plasticity from both sides. *Trends in Neurosciences* 26(12):662–70 doi:10.1016/j.tins.2003.09.014.

Sawaguchi, T. & Goldman-Rakic, P. S. 1991. 'D1 dopamine receptors in prefrontal cortex: involvement in working memory. *Science* 251(4996):947–50 doi:10.1126/science.1825731.

Schacter, D. L., *et al.* 1978. Richard Semon's theory of memory. *Journal of Verbal Learning and Verbal Behavior* 17(6):721–43 doi:10.1016/S0022-5371(78)90443-7.

Skaggs, W. E., *et al.* 1995. A model of the neural basis of the rat's sense of direction. *Advances in Neural Information Processing Systems 7,* edited by G. Tesauro *et al.* MIT Press, Massachusetts, USA 173–180. http://papers.nips.cc/paper/890-a-model-of-the-neural-basis-of-the-rats-sense-of-direction.pdf.

Standing, L, 1973. Learning 10000 pictures. *The Quarterly journal of experimental psychology, 25*(2):207–222.

Tang, Y. P., *et al.* 1999. Genetic enhancement of learning and memory in mice. *Nature* 401(6748):63–69 doi:10.1038/43432.

Wills, T. J., *et al.* 2005. Attractor dynamics in the hippocampal representation of the local environment. *Science* (New York, USA) 308(5723):873–76 doi:10.1126/science.1108905.

Chapter 5: Excitation and Inhibition

Albright, T. & Squire, L., editors. 2016. *The History of Neuroscience in Autobiography,* Volume 9. Academic Press, Massachusetts, USA.

Blair, E. A. & Erlanger, J. 1933. 'A comparison of the characteristics of axons through their individual electrical responses. *American Journal of Physiology* 106(3):524–64 doi:10.1152/ajplegacy.1933.106.3.524.

Börgers, C., *et al.* 2005. Background gamma rhythmicity and attention in cortical local circuits: a computational study. *Proceedings of the National Academy of Sciences of the United States of America* 102(19):7002–07 doi:10.1073/pnas.0502366102.

Brunel, N. 2000. Dynamics of sparsely connected networks of excitatory and inhibitory spiking neurons. *Journal of Computational Neuroscience* 8(3):183–208 doi:10.1023/A:1008925309027.

Fields, R. D. 2018. Do brain waves conduct neural activity like a
    symphony? *Scientific American* https://www.scientificamerican.
    com/article/do-brain-waves-conduct-neural-activity-like-a-
    symphony. Accessed 14 April 2020.
Florey, E. 1991. GABA: history and perspectives. *Canadian Journal of
    Physiology and Pharmacology* 69(7):1049–56 doi:10.1139/y91-156.
Fye, W. B., Ernst, W. & Weber E. *Clinical Cardiology* 23(9):709–10
    doi:10.1002/clc.4960230915.
Mainen, Z. F. & Sejnowski, T. J. 1995. Reliability of spike timing in
    neocortical neurons. *Science* 268(5216):1503–06, doi:10.1126/
    science.7770778.
Brown University. 2019. Neuroscientists discover neuron type that
    acts as brain's metronome: by keeping the brain in sync, these
    long-hypothesized but never-found neurons help rodents to
    detect subtle sensations. *ScienceDaily* https://www.sciencedaily.
    com/releases/2019/07/190718112415.htm. Accessed 14 April
    2020.
Poggio, G. F. & Viernstein, L. J. 1964. Time series analysis of impulse
    sequences of thalamic somatic sensory neurons. *Journal of
    Neurophysiology* 27(4):517–45 doi:10.1152/jn.1964.27.4.517.
Shadlen, M. N. & Newsome, W. T. 1994. Noise, neural codes and
    cortical organization. *Current Opinion in Neurobiology* 4(4):569–79
    doi:10.1016/0959-4388(94)90059-0.
Softky, W. R. & Koch, C. 1993. The highly irregular firing of cortical
    cells is inconsistent with temporal integration of random EPSPs.
    The Journal of Neuroscience 13(1):334–50 doi:10.1523/
    JNEUROSCI.13-01-00334.1993.
Stevens, C. F. & Zador, A. M. 1998. Input synchrony and the irregular
    firing of cortical neurons. *Nature Neuroscience* 1(3):210–17
    doi:10.1038/659.
Strawson, G. 1994. The impossibility of moral responsibility.
    *Philosophical Studies: An International Journal for Philosophy in the
    Analytic Tradition* 75(1/2):5–24 https://www.jstor.org/
    stable/4320507. JSTOR.
Tolhurst, D. J., *et al.* 1983. The statistical reliability of signals in single
    neurons in cat and monkey visual cortex. *Vision Research*
    23(8):775–85 doi:10.1016/0042-6989(83)90200-6.
Wehr, M. & Zador, A. M. 2003. Balanced inhibition underlies tuning
    and sharpens spike timing in auditory cortex. *Nature*
    426(6965):442–46 doi:10.1038/nature02116.

Chapter 6: Stages of Sight

Boden, M. A. 2006. *Mind as Machine: A History of Cognitive Science.* Clarendon Press, Oxford, UK.

Buckland, M. K. 2006. *Emanuel Goldberg and His Knowledge Machine.* Greenwood Publishing Group, Connecticut, USA.

Cadieu, C. F., *et al.* 2014. Deep neural networks rival the representation of primate IT cortex for core visual object recognition.' *PLoS Computational Biology* 10(12) doi:10.1371/journal.pcbi.1003963.

Fukushima, K. 1970. A Feature extractor for curvilinear patterns: a design suggested by the mammalian visual system.' *Kybernetik* 7(4):153–60 doi:10.1007/BF00571695.

Fukushima, K. 1980. Neocognitron: a self-organizing neural network model for a mechanism of pattern recognition unaffected by shift in position. *Biological Cybernetics* 36(4):193–202 doi:10.1007/BF00344251.

He, K., *et al.* 2015. Delving deep into rectifiers: surpassing human-level performance on ImageNet classification. ArXiv:1502.01852 [Cs] http://arxiv.org/abs/1502.01852.

Hubel, D. H. & Wiesel, T. N. 1962. Receptive fields, binocular interaction and functional architecture in the cat's visual cortex. *The Journal of Physiology* 160(1):106–154.2 www.ncbi.nlm.nih.gov/pmc/articles/PMC1359523/.

Hull, J. J. 1994. A database for handwritten text recognition research. *IEEE Computer Society* https://doi.org/10.1109/34.291440.

Husbands, P., *et al. An Interview with Oliver Selfridge.* The MIT Press, Massachusetts, USA. https://mitpress.universitypressscholarship.com/view/10.7551/mitpress/9780262083775.001.0001/upso-9780262083775-chapter-17. Accessed 14 April 2020.

Interview with Kunihiko Fukushima. 2015. CIS Oral History Project. IEEE.Tv https://ieeetv.ieee.org/video/interview-with-fukushima-2015. Accessed 14 Apr. 2020.

Khaligh-Razavi, S. M. & Kriegeskorte, N. 2014. Deep supervised, but not unsupervised, models may explain IT cortical representation.' *PLOS Computational Biology* 10(11):e1003915 doi:10.1371/journal.pcbi.1003915.

Krizhevsky, A., *et al.* 2017. ImageNet classification with deep convolutional neural networks. Association for Computing Machinery https://doi.org/10.1145/3065386.

LeCun, Y., *et al.* 1989. Backpropagation applied to handwritten zip code recognition. *Neural Computation* 1(4):541–51 doi:10.1162/neco.1989.1.4.541.

National Physical Laboratory. 1959. Mechanisation of thought processes; proceedings of a symposium held at the National Physical Laboratory on 24th, 25th, 26th and 27th November 1958. H. M. Stationery Office, London, UK.

Papert, S. A. 1966. The summer vision project. https://dspace.mit.edu/handle/1721.1/6125.

Squire, L. R., editor. 1998. *The History of Neuroscience in Autobiography, Volume 1.* Academic Press, Massachusetts, USA.

Uhr, L. 1963. Pattern recognition computers as models for form perception. *Psychological Bulletin* 60:40–73 doi:10.1037/h0048029.

Chapter 7: Cracking the Neural Code

Barlow, H. 2001. Redundancy reduction revisited. *Network* (Bristol, England) 12(3):241–53.

Barlow, H. B. 2012. Possible principles underlying the transformations of sensory messages. *Sensory Communication*, edited by Walter A. Rosenblith, The MIT Press, Massachusetts, USA. 216–34 doi:10.7551/mitpress/9780262518420.003.0013.

Barlow, H. B. 1972. Single units and sensation: a neuron doctrine for perceptual psychology? *Perception* 1(4):371–94 doi:10.1068/p010371.

Engl, E. & Attwell, D. 2015. Non-signalling energy use in the brain. *The Journal of Physiology* 593(16):3417–29 doi:10.1113/jphysiol.2014.282517.

Fairhall, A. L., *et al.* 2001. Efficiency and ambiguity in an adaptive neural code. *Nature* 412(6849):787–92 doi:10.1038/35090500.

Foster, M. 1870. The velocity of thought. *Nature* doi:10.1038/002002a0. Accessed 14 April 2020.

Gerovitch, S. 2004. *From Newspeak to Cyberspeak: A History of Soviet Cybernetics.* MIT Press, Massachusetts, USA.

Gross, C. G. 2002. Genealogy of the 'grandmother cell'. *The Neuroscientist: A Review Journal Bringing Neurobiology, Neurology and Psychiatry* 8(5):512–18 doi:10.1177/107385802237175.

Hodgkin, A. 1979. Edgar Douglas Adrian, Baron Adrian of Cambridge, 30 November 1889–4 August 1977. *Biographical Memoirs of Fellows of the Royal Society.* Royal Society, Great Britain. 25:1–73 doi:10.1098/rsbm.1979.0002.

Horgan, J. 2017. Profile of Claude Shannon, inventor of information theory. *Scientific American* Blog Network https://blogs. scientificamerican.com/cross-check/profile-of-claude-shannon-inventor-of-information-theory. Accessed 14 April 2020.

Husbands, P., *et al.* 2008. An interview with Horace Barlow. The MIT Press https://mitpress.universitypressscholarship.com/ view/10.7551/mitpress/9780262083775.001.0001/upso-9780262083775-chapter-18. Accessed 14 April 2020.

Joris, P. X., *et al.* 1998. Coincidence detection in the auditory system: 50 years after Jeffress. *Neuron* 21(6):1235–38 doi:10.1016/ s0896-6273(00)80643-1.

Lewicki, M. S. 2002. Efficient coding of natural sounds. *Nature Neuroscience* 5(4):356–63 doi:10.1038/nn831.

Perkel, D. H. 1968. Neural coding: a report based on an NRP work session organized by Theodore Holmes bullock and held on January 21–23, 1968. Neurosciences Research Program.

Smeds, L., *et al.* 2019. Paradoxical rules of spike train decoding revealed at the sensitivity limit of vision. *Neuron* 104(3):576–587. e11 doi:10.1016/j.neuron.2019.08.005.

Stein, R. B. 1967. The information capacity of nerve cells using a frequency code. *Biophysical Journal* 7(6):797–826 https://www. ncbi.nlm.nih.gov/pmc/articles/PMC1368193.

The Hospital Nursing Supplement. 1892. The Hospital 12(309):153–60 https://www.ncbi.nlm.nih.gov/pmc/articles/PMC5281805.

Von Foerster, H. 2013. *The Beginning of Heaven and Earth Has No Name: Seven Days with Second-Order Cybernetics.* Fordham University Press, New York, USA.

Chapter 8: Movement in Low Dimensions

Ashe, J. 2005. What is Coded in the Primary Motor Cortex ? *Motor Cortex in Voluntary Movements: A Distributed System for Distributed Functions.* CRC Press, Massachusetts, USA doi:10.1201/9780203503584.ch5.

Carr, L. 2012. The neural rhythms that move your body. *The Atlantic* www.theatlantic.com/health/archive/2012/06/the-neural-rhythms-that-move-your-body/258094.

Churchland, M. M., *et al.* 2010. Cortical preparatory activity: representation of movement or first cog in a dynamical machine? 68(3):387–400 doi:10.1016/j.neuron.2010.09.015.

Clar, S. A & Cianca, J. C. 1998. Intracranial tumour masquerading as cervical radiculopathy: a case study. *Archives of Physical Medicine and Rehabilitation* 79(10):1301–02 doi:10.1016/S0003-9993(98)90279-9.

Evarts, E. V. 1968. Relation of pyramidal tract activity to force exerted during voluntary movement. *Journal of Neurophysiology* 31(1):14–27 doi:10.1152/jn.1968.31.1.14.

Ferrier, D. 1876. *The Functions of the Brain*. Smith, Elder & Co, London. archive.org/details/functionsofbrain1876ferr.

Fetz, E. E. 1992. Are movement parameters recognizably coded in the activity of single neurons? *Behavioral and Brain Sciences* 15(4):679–90.

Finger, S., et al. 2009. *History of Neurology*. Elsevier, Amsterdam, Netherlands.

Georgopoulos, A. P. 1998. Interview with Apostolos P. Georgopoulos. *Journal of Cognitive Neuroscience* 10(5):657–61 doi:10.1162/089892998562951.

Kalaska, J. F. 2009. From intention to action: motor cortex and the control of reaching movements. *Advances in Experimental Medicine and Biology* 629:139–78 doi:10.1007/978-0-387-77064-2_8.

Kaufman, M. T., et al. 2014. Cortical activity in the null space: permitting preparation without movement. *Nature Neuroscience* 17(3):440–48 doi:10.1038/nn.3643.

Rioch, D. M. 1938. Certain aspects of the behavior of decorticate cats. *Psychiatry* 1(3):339–45 doi:10.1080/00332747.1938.11022202.

Shenoy, K. V., et al. 2013. Cortical control of arm movements: a dynamical systems perspective. *Annual Review of Neuroscience* 36:337–59 doi:10.1146/annurev-neuro-062111-150509.

Squire, L. R., editor. 2009. *The History of Neuroscience in Autobiography. Volume 6*. Oxford University Press, USA.

Taylor, C. S. R., & Gross, C. G. 2003. Twitches versus movements: a story of motor cortex. *The Neuroscientist: A Review Journal Bringing Neurobiology, Neurology and Psychiatry* 9(5):332–42 doi:10.1177/1073858403257037.

Venkataramanan, M. 2015. A chip in your brain can control a robotic arm. Welcome to BrainGate. *Wired UK* www.wired.co.uk/article/braingate.

Whishaw, I. Q. & Kolb, Bryan. 1983. Can male decorticate rats copulate? *Behavioral Neuroscience* 97(2):270–79 doi:10.1037/0735-7044.97.2.270.

Wickens, A. P. 2014. *A History of the Brain: From Stone Age Surgery to Modern Neuroscience*. Psychology Press, East Sussex, UK.

Chapter 9: From Structure to Function

Bassett, D.S. *et al.*, 2008. Hierarchical organization of human cortical networks in health and schizophrenia. *Journal of Neuroscience, 28*(37), 9239–9248.

Fornito, A., *et al.*, editors. 2016. Chapter 8 – Motifs, Small Worlds, and Network Economy. *Fundamentals of Brain Network Analysis.* Academic Press, London, UK. 257–301 doi:10.1016/B978-0-12-407908-3.00008-X.

Garcia-Lopez, P., *et al.* 2010. The histological slides and drawings of Cajal. *Frontiers in Neuroanatomy* 4 doi:10.3389/neuro.05.009.2010.

Griffa, A., *et al.* 2013. Structural connectomics in brain diseases. *NeuroImage* 80: 515–26 doi:10.1016/j.neuroimage.2013.04.056.

Hagmann, P., *et al.* 2007. Mapping human whole-brain structural networks with diffusion MRI. *PLOS ONE* 2(7):e597 doi:10.1371/journal.pone.0000597.

Heuvel, M. P. van den, & Sporns, Olaf. 2013. Network hubs in the human brain. *Trends in Cognitive Sciences* 17(12):683–96 doi:10.1016/j.tics.2013.09.012.

Humphries, M. D., *et al.* 2006. The brainstem reticular formation is a small-world, not scale-free, network. *Biological Sciences* 273(1585):503–11 doi:10.1098/rspb.2005.3354.

Marder, E. & Taylor, A. L. 2011. Multiple models to capture the variability in biological neurons and networks. *Nature Neuroscience* 14(2):133–38 doi:10.1038/nn.2735.

Milgram, S. 1967. The small world problem. *Psychology Today* 2:60–67.

Mohajerani, M. H. & Cherubini, E. 2006. Role of giant depolarizing potentials in shaping synaptic currents in the developing hippocampus. *Critical Reviews in Neurobiology* 18(1–2):13–23 doi:10.1615/critrevneurobiol.v18.i1-2.30.

Morrison, K. & Curto, C. 2019. Chapter 8 – Predicting Neural Network Dynamics via Graphical Analysis. *Algebraic and Combinatorial Computational Biology*, edited by Robeva, Raina & Macauley, M, Academic Press, London, UK. 241–77 doi:10.1016/B978-0-12-814066-6.00008-8.

Muldoon, S. F., *et al.* 2016. Stimulation-based control of dynamic brain networks. *PLOS Computational Biology* 12(9):e1005076 doi:10.1371/journal.pcbi.1005076.

Navlakha, S., *et al.* 2018. Network design and the brain. *Trends in Cognitive Sciences*, 22:64–78 doi:10.1016/j.tics.2017.09.012.

Servick, K. 2019. This physicist is trying to make sense of the brain's tangled networks. *Science | AAAS*, www.sciencemag.org/ news/2019/04/physicist-trying-make-sense-brain-s-tangled-networks.

Sporns, O. *et al.* 2004. Organization, development and function of complex brain networks. *Trends in Cognitive Sciences* 8(9):418–25 doi:10.1016/j.tics.2004.07.008.

Sporns, O. *et al.* 2005. The human connectome: a structural description of the human brain. *PLOS Computational Biology* 1(4):e42 doi:10.1371/journal.pcbi.0010042.

Squire, L. R. & Albright, T. D. editors. 2008. *The History of Neuroscience in Autobiography Volume 9*. Oxford University Press, New York, USA.

Squire, L. R. & Albright, T. D. editors. 2008. *The History of Neuroscience in Autobiography Volume 10*. Oxford University Press, New York, USA.

Tau, G. Z. & Peterson, B. S. 2010. Normal development of brain circuits. *Neuropsychopharmacology* 35(1):147–68 doi:10.1038/ npp.2009.115.

Tijms, B. M. *et al.*, 2013. Single-subject grey matter graphs in Alzheimer's disease. *PloS one, 8*(3), e58921.

Towlson, E. K., *et al.* 2013. The rich club of the *C. Elegans* neuronal connectome. *The Journal of Neuroscience* 33(15):6380–87 doi:10.1523/JNEUROSCI.3784-12.2013.

Watts, D. J. & Strogatz, S. H. 1998. Collective dynamics of 'small-world' networks. *Nature* 393(6684):440–42 doi:10.1038/30918.

Chapter 10: Making Rational Decisions

Stix, G. 2014. A conversation with Dora Angelaki. *Cold Spring Harbor Symposia on Quantitative Biology* 79:255–57 doi:10.1101/ sqb.2014.79.02.

Adams, W. J., *et al.* 2004. Experience can change the 'light-from-above' prior. *Nature Neuroscience* 7(10):1057–58 doi:10.1038/nn1312.

Aitchison, L., *et al.* 2015. Doubly Bayesian analysis of confidence in perceptual decision-making. *PLoS Computational Biology* 11(10) doi:10.1371/journal.pcbi.1004519.

Anderson, J. R. 1991. Is human cognition adaptive? *Behavioral and Brain Sciences* 14(3):471–85 doi:10.1017/S0140525X00070801.

Bowers, J. S. & Davis, C. J. 2012. Bayesian Just-so Stories in psychology and neuroscience. *Psychological Bulletin* 138(3):389–414 doi:10.1037/a0026450.

Cardano, G. 2002. *The Book of My Life*. New York Review Books, USA.

Curry, R. E. 1972. A Bayesian model for visual space perception. *NASSP* 281:187 https://ui.adsabs.harvard.edu/abs/1972NASSP.281..187C/abstract.

Fetsch, C. R., *et al.* 2009. Dynamic reweighting of visual and vestibular cues during self-motion perception. *Journal of Neuroscience* 29(49):15601–12 doi:10.1523/JNEUROSCI.2574-09.2009.

Fisher, R. A. & Russell, E. J. 1922. On the mathematical foundations of theoretical statistics. *Philosophical Transactions of the Royal Society of London. Series A, Containing Papers of a Mathematical or Physical Character* 222(594–604):309–68 doi:10.1098/rsta.1922.0009.

Gillies, D. A. 1987. Was Bayes a Bayesian? *Historia Mathematica* 14(4):325–46 doi:10.1016/0315-0860(87)90065-6.

Gorroochurn, P. 2016. *Classic Topics on the History of Modern Mathematical Statistics: From Laplace to More Recent Times*. John Wiley & Sons, New Jersey, USA.

Helmholtz, H. von & Southall, J. P. C. 2005. *Treatise on Physiological Optics*. Dover Publications, New York, USA.

Jaynes, E. T. 2003. *Probability Theory: The Logic of Science: Principles and Elementary Applications Vol 1*. Edited by G. Larry Bretthorst, Cambridge University Press, New York, USA.

Koenigsberger, L. 1906. *Hermann von Helmholtz*. Clarendon Press, Oxford, UK.

Mamassian, P. 2008. Ambiguities and conventions in the perception of visual art. *Vision Research* 48(20):2143–53 doi:10.1016/j.visres.2008.06.010.

Moreno-Bote, R., *et al.* 2011. Bayesian sampling in visual perception. *Proceedings of the National Academy of Sciences* 108(30):12491–96 doi:10.1073/pnas.1101430108.

Seriès, P. & Seitz, A. R. 2013. Learning what to expect (in visual perception). *Frontiers in Human Neuroscience* 7:668 doi:10.3389/fnhum.2013.00668.

Stigler, S. M. 1982. Thomas Bayes's Bayesian inference. *Journal of the Royal Statistical Society*. Series A (General) 145(2):250–58 doi:10.2307/2981538. JSTOR.

Vilares, I. & Kording, K. 2011. Bayesian Models: the structure of the world, uncertainty, behavior, and the brain. *Annals of the New York*

Academy of Sciences 1224(1):22–39 doi:10.1111/j.1749-6632.2011.05965.x.

Weiss, Y., *et al.* 2002. Motion illusions as optimal percepts. *Nature Neuroscience* 5(6):598–604 doi:10.1038/nn0602-858.

Chapter 11: How Rewards Guide Actions

Bellman, R. 1984. *Eye of the Hurricane.* World Scientific, Singapore.

Bellman, R. E. 1954. The theory of dynamic programming. www.rand.org/pubs/papers/P550.html.

Bergen, M. 2016. Google has found a business model for its most advanced artificial intelligence. *Vox* www.vox.com/2016/7/19/12231776/google-energy-deepmind-ai-data-centers.

Mnih, V., et al. 2013. Playing Atari with deep reinforcement learning. ArXiv:1312.5602 [Cs], http://arxiv.org/abs/1312.5602.

Redish, A. D. 2004. Addiction as a computational process gone awry. *Science* (New York) 306(5703):1944–47 doi:10.1126/science.1102384.

Rescorla, R. A. & Wagner, A. 1972. A theory of Pavlovian Conditioning: variations in the effectiveness of reinforcement and nonreinforcement. *Classical Conditioning II: Current Research and Theory* 2

Schultz, W., Dayan, P., *et al.* 1997. A neural substrate of prediction and reward. *Science* (New York) 275(5306):1593–99 doi:10.1126/science.275.5306.1593.

Schultz, W., Apicella, P., *et al.* 1993. Responses of monkey dopamine neurons to reward and conditioned stimuli during successive steps of learning a delayed response task. *The Journal of Neuroscience: The Official Journal of the Society for Neuroscience* 13(3):900–13.

Sejnowski, T. J. 2018. *The Deep Learning Revolution.* MIT Press, Massachusetts, USA.

Specter, Michael. 2014. Drool. *The New Yorker* www.newyorker.com/magazine/2014/11/24/drool. Accessed 14 April 2020.

Story, G. W., et al. 2014. Does temporal discounting explain unhealthy behavior? a systematic review and reinforcement learning perspective. *Frontiers in Behavioral Neuroscience* 8 doi:10.3389/fnbeh.2014.00076.

Sutton, R. S. 1988. Learning to predict by the methods of temporal differences. *Machine Learning* 3(1):9–44 doi:10.1007/BF00115009.

Chapter 12: Grand Unified Theories of the Brain

Anderson, M. L. & Chemero, T. 2013. The Problem with brain GUTs: conflation of different senses of 'prediction' threatens metaphysical disaster. *The Behavioral and Brain Sciences* 36(3):204–05 doi:10.1017/S0140525X1200221X.

Buxhoeveden, D. P. & Casanova, Manuel F. 2002. The minicolumn hypothesis in neuroscience. *Brain* 125(5):935–51 doi:10.1093/brain/awf110.

Clark, J. 2014. Meet the man building an AI that mimics our neocortex – and could kill off neural networks. www.theregister.co.uk/2014/03/29/hawkins_ai_feature.

Eliasmith, C., *et al.* 2012. A large-scale model of the functioning brain. *Science* 338(6111):1202–05 doi:10.1126/science.1225266.

Fridman, L. 2019. Jeff Hawkins: Thousand Brains Theory of intelligence. https://lexfridman.com/jeff-hawkins. Accessed 14 Apr. 2020.

Friston, K. 2019. A free energy principle for a particular physics. ArXiv:1906.10184 [q-Bio] http://arxiv.org/abs/1906.10184.

Friston, K. 2010. The free-energy principle: a unified brain theory? *Nature Reviews Neuroscience* 11(2):127–38 doi:10.1038/nrn2787.

Friston, K., Fortier, M. & Friedman, D. A. 2018. Of woodlice and men: a Bayesian account of cognition, life and consciousness. An interview with Karl Friston. *ALIUS Bulletin*, 2:17–43.

Hawkins, J., *et al.* 2019. A framework for intelligence and cortical function based on grid cells in the neocortex. *Frontiers in Neural Circuits* 12 doi:10.3389/fncir.2018.00121.

Heilbron, M. & Chait, M. 2018. Great expectations: is there evidence for predictive coding in auditory cortex? *Neuroscience* 389:54–73 doi:10.1016/j.neuroscience.2017.07.061.

Metz, C. 2018. Jeff Hawkins is finally ready to explain his brain research. *The New York Times* www.nytimes.com/2018/10/14/technology/jeff-hawkins-brain-research.html.

Michel, M., *et al.* 2018. An informal internet survey on the current state of consciousness science. *Frontiers in Psychology* 9 doi:10.3389/fpsyg.2018.02134.

Nanopoulos, D. V. 1979. Protons are not forever. *High-Energy Physics in the Einstein Centennial Year*, edited by Arnold Perlmutter et al. Springer US. 91–114 doi:10.1007/978-1-4613-3024-0_4.

Rao, R. P. & Ballard, D. H., 1999. Predictive coding in the visual cortex: a functional interpretation of some extra-classical receptive-field effects. *Nature Neuroscience, 2*(1), pp.79–87.

Raviv, S. The Genius Neuroscientist Who Might Hold the Key to
    True AI. Wired, https://www.wired.com/story/karl-friston-free-
    energy-principle-artificial-intelligence/. Accessed 14 Apr. 2020.
Rummell, B. P., *et al.* 2016. Attenuation of responses to self-generated
    sounds in auditory cortical neurons. *Journal of Neuroscience*
    36(47):12010–26 doi:10.1523/JNEUROSCI.1564-16.2016.
Simonite, T. 2015. IBM tests mobile computing pioneer's controversial
    brain algorithms. *MIT Technology Review* www.technologyreview.
    com/2015/04/08/11480/ibm-tests-mobile-computing-pioneers-
    controversial-brain-algorithms.
Tononi, G., *et al.* 2016. Integrated information theory: from
    consciousness to its physical substrate. *Nature Reviews Neuroscience*
    17(7):450–61 doi:10.1038/nrn.2016.44.

Index

Page numbers in **bold** indicate figures.